Praise for Kirkpatrick Sale's previous books:

for *The Fire of His Genius: Robert Fulton and the American Dream*

"Excellent. . . . A truly marvelous biography of a misunderstood American hero."–DOUGLAS BRINKLEY, TULANE UNIVERSITY

"Superb, beautifully written. . . . An informative, moving story that personalizes the relatively obscure life of a self-taught tinkerer who had a genius for self-promotion and exploiting the discoveries of others."
–LIBRARY JOURNAL

"Kirkpatrick Sale is one of those writers whose pen will always set the imagination alight regardless of his topic."*–FOURTH WORLD REVIEW*

for *Rebels against the Future: The Luddites and Their War on the Industrial Revolution: Lessons for the Computer Age*

"The message of Sale's *Rebels against the Future* is one of the most urgent of that in any book this century."*–MINNEAPOLIS STAR TRIBUNE*

"If Sale's book serves no other purpose than getting us to ask ourselves whether technology serves us or we serve technology, it will be worth reading."*–WASHINGTON POST*

for *The Green Revolution: The American Environmental Movement, 1962–1992*

"I doubt that any 108 pages in the English language come close to educating the reader about the environment—its despoilers and its defenders—as well as Kirkpatrick Sale's small volume. *The Green Revolution* is a tour de force. It is brief but comprehensive, analytical but readable, insightful, factual, current but historical and futu   " RALPH NADER

"A virtual primer on current environmental politics. . . . impassioned in its advocacy and provocative in its analysis."–RICHARD WHITE, STANFORD UNIVERSITY

for _The Conquest of Paradise: Christopher Columbus and the Columbian Legacy_

"Sale is an invaluable guide and teacher. . . . The breadth of his scholarship and the skillful mix of historical fact, conviction, and style makes this an absorbing book."–_ECONOMIST_

"Brilliant . . . well-researched, insightful, and fascinating in both its portrait of the Great Discoverer's complex and tragic character and that of the age in which he lived."–_SAN FRANCISCO CHRONICLE_

"Scholarly, ambitious, meticulous, and angry."–_WASHINGTON POST BOOK WORLD_

"The author's lucid, vigorous prose is a delight, whether he is narrating perilous passages at sea or steering us through the hazards of learned disputes."–_NEW YORKER_

"A work of haunting attraction, of splendid magnitude, illuminating scholarship, and compelling strength and imagination. I find it a stunning achievement."–JOSEPH HELLER

"Acute and tough-minded . . . fresh and highly readable."–LARRY MCMURTRY

for _Dwellers in the Land: The Bioregional Vision_

"_Dwellers in the Land_ is at once an alarm, a directive, and a tonic. One could hardly ask for more."–_KANSAS CITY STAR_

AFTER EDEN

AFTER EDEN

THE EVOLUTION OF

EDEN

HUMAN DOMINATION

KIRKPATRICK SALE

Duke University Press   Durham and London   2006

© 2006 Duke University Press
All rights reserved.
Printed in the United States of America on acid-free paper ∞
Designed by Heather Hensley
Typeset in Adobe Garamond Pro by Tseng Information Systems, Inc.
Library of Congress Cataloging-in-Publication Data appear
on the last printed page of this book.

There were at least two broad and overlapping epochs
in prehistory: one in which our ancestors confronted the
world, for the most part, as potential prey, and another
in which they took their place among the predators which
had for so long oppressed them. The transition from one
status to the other . . . had to be the single greatest advance
in human evolution. — **Barbara Ehrenreich,** *Blood Rites*

Man is the most dominant animal that has ever appeared
on earth. — **Charles Darwin,** *The Descent of Man*

CONTENTS

Introduction

Think of it: ours is the only species in all the long period of life on earth that has ever spread around the entire world, occupying every continent and nearly every island, effectively subjugating all the animals and plants and natural systems it has encountered and over a relatively short period of time establishing itself in vast numbers as the single most dominant species of all.

It is a wonder that this single fact seems not to have imprinted itself powerfully upon the consciousness of us all, filling us with daily amazement, or bewilderment, or awe. Nor has it been the bedrock of religions, regarded as a phenomenon of human triumph worthy of praise and gratitude and homage. Nor the preoccupation of scholars, recounting how this marvelous and unique transformation has taken place and investigating *why* it is that humans, and only humans, are everywhere. Nor the inspiration of the arts, celebrating the immensity of this achievement and the treasures it has devolved upon us.

Instead, we take it for granted. That is the way things are, and why not?

But in the preceding two million years none of the earlier hominid species did anything like that. Their effect on the environment was very limited, as were their numbers at any one place, and they lived in settings confined in space and impact, not terribly different from other mammalian species except that they had stone tools and intimate social groupings. *Homo erectus*, it is true, did spread out of Africa several times, into the Levant, the Caucases, and southern Asia and China, but in only modest numbers and with limited technologies insufficient for widespread conquest and dominion.

It was not until our species, *Homo sapiens* we pridefully call it, came upon the earth and began to use its bigger and more pleated brain and its more complex technologies and its more elaborate social networks that any creature was able regularly to control an increasing number of the other elements of the natural world for its own use and pleasure. And it was not until our species that any creature developed the psychological perceptions, and then the behavioral manifestations, that allowed it to stand *distanced* from that world—indeed, in many ways opposed to it—so that it was able to expand its control and dominion in a totally unprecedented way, justified in the name of survival.

Those uncountable eons of hominid evolution, and the transformation at the heart of it wrought by our species, may have been what was in the tribal memory of the ancient Hebrews when they set down the story of humankind. They began it with Eden, where humans were given a garden "to dress it and keep it," living compatibly with "every beast of the field and every fowl of the air" and foraging from trees "pleasant to the sight, and good for food"—in other words, the world of pre-Sapiens humanity, in effective harmony with nature. But after the transgression of Adam and Eve, when humans were turned out of paradise and forced to wrest a living from the cursed earth, and then still later when the Lord decreed that "every moving thing that liveth shall be meat for you" and "the dread of you shall be upon every beast of the earth," a new and adversarial role for humankind among the species of the earth was set. After Eden.

The evolution of human dominance, then, begins with the evolution of *Homo sapiens* into a creature that could kill "every beast of the earth" and thus in time conquer and transform the world. After Eden. That is the story of this book.

Consider the extent of our domination. Modern humans, now numbering six billion and predicted to go to ten billion, have left not one ecosystem on the surface of the earth free of pervasive human influence, transforming more than half the land on the planet for their own use (a quarter for farming and forestry, a quarter for pasture, 3 per cent for industry, housing, and transport), consuming more than 40 per cent of the total photosynthetic productivity of the sun, using 55 per cent of the world's freshwater, controlling and regulating two-thirds of all the rivers and streams, and consuming a vast variety of plant, animal, and mineral resources, often to depletion, at a pace that is estimated not to be sustainable for more than fifty years. In the words of the *Atlas of Population and Environment*, compiled by the American Academy for the Advancement of Science in 2000, "Like volcanoes and earthquakes, humans have become a force in nature." Only more so—humans now produce about three thousand times more heat energy on average than the world's volcanoes—and on a worldwide scale without precedence.

It is this extraordinary dominance by one single bipedal species that has brought us to the present imperilment of the earth, including the extinction of species, the destruction of ecosystems, the alteration of climate, the pollution of waters and soils, the exhaustion of fisheries, the elimination of forests, the spread of deserts, and the disruption of the atmosphere. There is some dispute about when the ecological catastrophe as a result of all this is likely to hit us full force, and in what ghastly form, but it is no exaggeration to say that the undeniable scientific and informed consensus is that if western civilization continues its reckless policies and practices toward the earth we are headed toward *ecocide*.

I need not belabor the point. The threat is real and well recognized. Fourteen years ago the Union of Concerned Scientists, in a statement endorsed by seventeen hundred members of various national academies of science around the globe, issued a "World Scientists' Warning to Humanity," stating that if present rates of environmental assault were not halted the planet would be so "irretrievably mutilated" as to "be unable to sustain life in the manner that we know it." In the years since then, the assault has gotten far worse: according to a landmark Ecosystem Millennium Assessment issued by 1,360 scientists in March 2005, the most comprehensive environmental review ever done, two-thirds of the natural world that supports life on earth is being degraded by human pressure, and "human activity is putting such a strain on the natural functions of Earth that the ability of the planet's ecosystems to sustain future generations can no longer be taken for granted." This is serious stuff: our domination threatens our survival. As the eminent biologist Edward O. Wilson, no alarmist, has asserted, "the appropriation of productive land—the ecological footprint—is already too large for the planet to sustain" and has likely stressed the earth beyond "its ability to regenerate."

Given the ultimate cruciality of this issue, I want to say that the most important questions for our time are without doubt: How did we get into this predicament? How did humankind develop the attitude that it was right and good for it to dominate nature as much as possible, to regard the earth and its creatures and its habitats as there for unconstrained human consumption and manipulation? And how can humankind possibly alter that perception, learn to live in balance with nature as we once did during the long ages that made up 99 per cent of the time humans have been on earth?

Luckily, we are aided in this search by the recent developments in paleontology and paleoanthropology in just the last decade or so—an increasing number of carefully excavated sites and new technologies of analysis and dating (CAT scans, DNA extraction, argon dating, electron spin resonance, and much more)—that enable a fairly reliable picture of our long prehistory to emerge. Not that everything is resolved and certain, mind you, or all controversies dispelled. But enough evidence has emerged in recent years that we can begin to get a handle on some of the larger questions of our Stone Age heritage that had previously been evasive; as the anthropologists Alison Brooks and Rick Potts have recently put it, "With the help of the new technologies and an expanding data base, we are now in a better place to begin to answer these questions than ever before." And we *need* to answer them more than ever before.

I should say at the start that in investigating this period I bring a somewhat unusual perspective. Though I have spent much of my life reading randomly about the Stone Age, and in the last six years reading about it intensively for this book, I am not a paleoanthropologist or paleoarcheologist, not indeed a scientist at all. I am a historian. Technically the Stone Age is prehistory, since "history" is deemed to begin only with writing and that does not emerge until several thousands of years later, but in the truest sense all that makes up the human story, from maybe six million years ago to the present, is history. And so I come to the study of the ancient past not as a specialist in paleontology but as a curious historian, searching to find out how humans behaved those eons ago, particularly how the species *Homo sapiens* evolved and developed. There are no dusty tomes or yellowed records here, of course, as there have been for most of my work on the past, but there are fossils, fossils of all kinds, and they can tell an amazingly complete and complex story. And everything I have to say here, every argument or theory, is based on the fossil record as we know it today.

Oftentimes, it is true, there are different ways to interpret the fossil record and different specialists can arrive at different stories; and when that happens I find that as an outsider, as it were, as a generalist historian, I can be more objective than the partisan scientists pushing one view over another. Sometimes I come down more firmly on one side, while giving a full hearing to the other, and sometimes I find that the record does not convincingly justify either interpretation, and I can say so because I don't have any particular handaxe to grind. Moreover, I can freely move from one academic specialty to another, at least within the limits of my competence, and I do not have to approach an area as only an anthropologist or an archeologist, only a specialist in radiocarbon dating or cave art or Neandertal society. I do mostly have to rely on those specialists for the basic information from which to derive my history, and I carefully credit each one that I have used—the Source Notes at the end of the book list every article and book that I have consulted—but I am not committed necessarily to their points of view and when I deviate I explain how and why. The result, I feel, allows me a freedom to roam over the whole paleological record and the many disciplines involved in its production with an objective and creative perspective that only an outsider with a historical analysis could bring.

Let me give an example, having to do with the famous carvings and cave paintings that marked the art of European *Homo sapiens* from about 35,000 years ago down to the end of the Ice Age around 10,000 years ago. Much has been written about it, from the points of view of many different disciplines, but almost always with the assumption that because so much of it is powerful and beautiful to our contemporary eyes, it was so to the eyes of those ancients that created it. Not only that, but the art had to have been part of a virtuous, perhaps a spiritual, process of some sort, the product of people who had an intimate and meaningful relationship with the kinds of animals represented. And it came about for essentially benign reasons: as magical symbols used for initiation or other tribal ceremonies, or as ornaments of beauty to celebrate the animals on which the bands depended for their sustenance. Or, as the poet Gary Snyder has put it, they were "produced not for practical use but apparently for magic or beauty."

My interpretation of this art, in contrast, has nothing to do with benignity or esthetics. I believe it is a very practical, indeed vital, product of peoples who were in a severe crisis, facing a serious threat to their very survival, and who developed this extraordinary means of ritual and magic to allow them to get through this exigency. In this I agree with John Pfeiffer, another non-scientist, who argues in his study of what he calls the "creative explosion" that art arose in response to "a threat of survival of the species." But he does not specify what this threat might be or why it should lead to the crafting and depicting of large animals, something that had never been done in all the previous 130,000 years of Sapiens existence.

The threat, it seems clear to me, was a decline in the number of animals available for the hunt, increased competition for those left, and the danger of widespread starvation. And it arose because of two phenomena, occurring at about the time of the origin of art, that have only recently been thoroughly investigated and as far as I know never linked.

The first was a sharp drop in temperature in Europe after 35,000 years ago as the subcontinent began to enter a "full glacial" period that was to last until about 10,000 years ago—the exact era of artistic achievement. The cold, accompanied by a decrease in precipitation and the southward movement of the polar ice sheet (eventually resting at the middle of modern Germany), meant a shorter and drier growing season for vegetation and a decreased range for animals. Game—in particular arctic-adapted animals like mammoths, reindeer, and horses—and the people who depended on them gradually moved south, into regions of relative relief, particularly the river valleys of the Russian plain and southwestern France and Cantabrian Spain. But some species declined in numbers and body size and some did not survive the continuing cold—certain kinds of rabbits and deer and varieties of elephant, rhinoceros, and hippopotamus became extinct before 30,000 years ago—and the pressure on the hunting populations must have been severe.

Increased populations in the favored refuge regions would of course have increased competition for the dwindling number of animals, but that was greatly aggravated by the second phenomenon, a huge migration of Sapiens from central Asia, probably forced off the open steppe by the increasing cold and aridity. This has been discovered only recently by the population geneticist Spencer Wells, who traced Y-chromosome markers from residents of Asia that are present in the great majority of European peoples, and he suggests that this horde traveled across the mammoth steppe of Russia and the Ukraine, following migrating game herds, then appeared "on the scene so suddenly, around 35,000 years ago," along the major rivers of central and southwestern Europe.

It is not hard to see that under these extreme twin pressures the peoples of Europe would have turned to intensified forms of ritual to assure an adequate supply of meat, and a form of hunting magic that was enhanced by small, hand-held figurines and realistic art set deep in caves would have been among the most powerful. I do not think it an accident that the great majority of the painted caves of Europe are found in the Southwest, where the population densities were the greatest and the stress of success in the hunt would have been the most acute.

Thus by connecting a few of the dots provided me by scientific researchers—about herd sizes, extinctions, climate, genetics, and so on—I am able to suggest an answer not only to the vexing question of *why* humans created art but, perhaps more important, why they did it at just this time. That is how a historian, or more accurately a generalist, might see it.

Indeed, the whole idea of this book, embedded in an account of what I see as the origin of serious human hunting, and thus of human dominance, is an original one, not part of any previous paleological school of thought. I base it on evidence of critical climate change at about 71,000 years ago, provided by one group of paleoclimatologists, on evidence of the first stone weaponry at about the same time uncovered by another group of paleoarcheologists, and on evidence of other technical and behavioral innovations in human culture that followed, from the work of still other paleoscientists. I trace the evolution of this culture from southern Africa, the domain of one group of specialists, through Africa to the Mediterranean Levant, the domain of other specialists, and into Europe, the domain of still others. Because I can stand on the shoulders of this wide array of scholars, I can sometimes see farther, or at least differently, offering my own take on one aspect of human evolution that has, so far as I know, never been treated at this length and with this detail.

The first chapter of what follows takes us back to that 71,000-year-old climate crisis and how humans responded to it with means that changed human culture, and the natural order, forever. I have tried to recreate in some detail the character of this Sapiens culture as it developed in the struggle for existence during these difficult eons, for the most part using evidence from sites in southern Africa, showing that far from being innate, the will to domination was developed out of this crisis and the response to it of big-game hunting. Next I follow this successful culture in the period between 55,000 and 20,000 years ago as it spread out of Africa into the Levant and then Europe, expanding its techniques of domination and at the same time creating new means of dealing with the mental and social tensions this causes. The next chapter carries the story on for another 10,000 years or so, when another climatic crisis caused an intensification of the hunting lifeways of Sapiens society, particularly in Europe, culminating in widespread big-game extinctions there and elsewhere in the world. That period ends with the invention of agriculture and the domestication of both plant and animal species, the ultimate form of domination, but entirely of a pattern with the previous Sapiens experience.

In the last chapter I present the alternative to the culture of the Sapiens: the culture of the humans that preceded them, *Homo erectus*, who were the most successful of the various hominid species since we split from the chimpanzees six million years ago. Their span on earth lasted nearly two million years—more than ten times as long as the Sapiens'—and I suggest that their duration and achievement had more than a little to do with the way they perceived nature, quite differently from the Sapiens. The fossil record indicates little of the adversarial relationship with other creatures that existed at the core of Sapiens hunting society, and permits the conclusion that they must have lived in a deep, permeating bond with the natural world that the philosopher Owen Barfield has called "original participation," a "primal unity of mind and nature." And I propose that there are ways for us today to come to an appreciation of the Erectus consciousness, for something like it lingers on in various tribal societies on the fringes of civilization, in the core of such religions as Hinduism and Buddhism, even in certain parts of the worldwide environmental movement—and in our still-extant primal selves, for we experienced it for 72,000 generations of humanness.

And it might be that consciousness that will best save us from carrying out the Sapiens imperatives as we are doing today.

ONE

The Dawn of Modern Culture

70,000–50,000 YEARS AGO

A fierce and sudden volcanic winter descended upon the earth sometime around 71,000 years ago, when Mt. Toba, an enormous volcano on the island we know as Sumatra, in the Andaman Sea, exploded in the largest surface eruption that the earth has known for the past 400 million years. The mountain's ash and debris shot at least twenty miles into the air and were eventually scattered worldwide: 480 square miles of ash, it is reckoned, settled over the earth, burying all of the Pacific islands and the Indian subcontinent under a coat of as much as 10 inches of heavy sediments and darkening the skies around the globe with more than an incredible 1.1 billion tons of stratospheric dust and sulfuric acid aerosols. All that remained behind was a huge caldera, the largest natural lake in southeast Asia, 60 miles long and 36 miles wide.

Temperatures plunged. Michael Rampino, a geologist at New York University, has figured that the drop must have been about 15 to 25 degrees Fahrenheit worldwide, maybe 75 degrees in higher latitudes, wiping out many plants in the tropics and as much as two-thirds of them in some temperate climates, and resulting in summers at least 15 degrees cooler and winters even worse. Minute ash particles raining from the sky would have penetrated the lungs of many animals, impairing breathing, and settled in the feathers of birds, making flight impossible. Ice core records show that the sulfuric acid haze persisted in the atmosphere for six years, reflecting the sun's rays and keeping the earth in a perpetual winter. It was followed by a severe mini–ice age that lasted for as long as a thousand years, probably the coldest period of any during the final 60,000-year Ice Age that it ushered in. Northern forests became treeless moss- and lichen-covered tundra over permanently frozen earth or semiarid grasslands dotted with stunted shrubs, while tropical rainforests turned into dry open savanna, and grasslands into wind-swept deserts.

The results for the hominid species then on earth would have been catastrophic. Some remnant populations of *Homo erectus* were living in Indonesia and China, and a certain number of those not in the immediate vicinity of the ignimbrite lava flow would have survived, especially in the more tropical areas, but their populations, also affected by ash particles and now dependent mostly on a diminished animal supply and cold-water shellfish, would have dropped severely. The Neandertals of Europe and the Levant, with bodies already adapted to the cold and millennia of experience with cold-weather survival, would have fared better, but their food supplies would have dwindled too and the northern populations were probably hit far harder than the Mediterranean ones. And in Africa, where modern *Homo sapiens* populations lived, only those in a few favorable pockets not made unlivable by the extremely cold and dry conditions—such as in coastal southern and eastern regions and along the Mediterranean, where water was available and marine food supplies adequate—could have resisted the volcanic winter.

An increasing number of new fossil finds suggest that it was at this point, and most likely in response to these sudden extreme conditions, that the surviving Sapiens populations began to develop, or more aggressively adapt, both cultural and psychological mechanisms that turned them into fully modern beings.

Homo sapiens had been modern in *body* for maybe 90,000 years by then: the earliest fossils of what are called "anatomically modern humans" have been found at some two dozen sites in Africa and two in the Levant (south-western Asia but then geographically and climatologically an extension of Africa), the earliest at the Kibbish Formation in Ethiopia, dated to about 195,000 years ago. Whether this speciation process creating *Homo sapiens* was a relatively rapid event, as some paleontologists of the Stephen Jay Gould school argue, or a more gradual development out of *Homo erectus* and what is sometimes called "archaic Homo sapiens," as I am inclined to think the record shows, there is no doubt that some time around 150,000 years ago modern humans, fully erect, gracile in form and face, existed in Africa, with braincases in the range of 1,400–1,550 cubic centimeters (living human brains average about 1,350–1,400 cc.).[1]

But it has been difficult to determine if these early Sapiens were modern in *culture*—that is to say, in day-to-day behavior, in artifacts, in social units that would be generally recognizable to us today—or rather still used the tools and thought the thoughts of their ancestors. The general record (which is not very replete in Africa and difficult to date in the range of 250–127,000 years ago) indicates that for the most part the kinds of tools associated with these modern people were old-fashioned ones that had been in use, both in Africa and Europe, for the preceding 100,000 years. In a few places, it is true, as recent discoveries are showing, there were flashes of modern culture of the kind that would come to characterize the later fossil record. In Katanda, for example, a site in the eastern Congo on the Semliki River, eight whole or par-tial barbed stone harpoons and four worked bone artifacts have been dated to 90–80,000 years ago (though with a range of probable error that could extend the most recent date forward to 71,000 years), providing the first clear evidence of Africans fishing in a concerted way and one of the earliest ex-amples of the use of bone, and not just stone, to create weaponry. (Two earlier sites, Mumbwa and Broken Hill caves in Zambia, have bone fragments "puta-tively," but not convincingly, suggested as tools.) And in two sites in Tanzania, pieces of obsidian dated to 130,000 years ago have been found that originally came from the Rift Valley in Kenya, nearly 200 miles away, suggesting either a long-distance trading network or long-distance transport of valuable ma-terials, both hallmarks of the later modern culture. But these exceptions only serve to underscore that for the most part these early Sapiens were a people, as the anthropologist Richard Klein of Stanford University has put it, "similar to their precursors but very different from their successors."

Then, about 70,000 years ago, something changes. A modern culture begins to emerge.

It is my thesis that under the pressure of trying to survive the sudden and harsh volcanic winter, these Sapiens had to expand their means of wresting a living from nature in a dramatic way, leading to the kind of practices that would afterwards mark the species for the rest of the Stone Age—and indeed, in most ways, down to the present. Or, to put it another way, it was those humans who either developed or adapted in this modern way, with a culture centered on the regular hunting of a great variety of species, crafting better tools, creating new weapons from a wider set of materials, and developing rituals for tribal cohesion, who were able to survive in these harsh conditions. It is after all a basic principle of anthropology, as Robert Foley of Cambridge University puts it, that "ecological conditions . . . provide the basis for evolution in social behaviour," and these were ecological conditions in the extreme.

We can imagine them, several dozen clans of Sapiens who lived in deep caves along the coast of southern Africa, looking up with fear and astonishment as the huge gray clouds began to fill the blue sky, blocking out the sun and turning the warm day suddenly cold. We can sense their bewilderment and confusion, perhaps their anger, in the days and weeks afterward, as the gray clouds persisted, unlike any weather they had known before, and the temperatures kept dropping, vegetation began to wilt and die all around them, and many of the animals they depended upon seemed either to have migrated away or to have become victims of the falling ash and unusual cold.[2]

There are more than twenty caves and rock shelters along the southern coast that were occupied as the volcanic winter began, stretching in a 1,200-mile arc from the Steenbokfontein Cave on the Atlantic Ocean, some 200 miles north of the Cape of Good Hope, around the tip of the continent into the Indian Ocean, and up the east coast as far as Border Cave near modern-day Swaziland. This would have been a particularly advantageous place for human and animal survival, despite its low latitude, because the Cape region

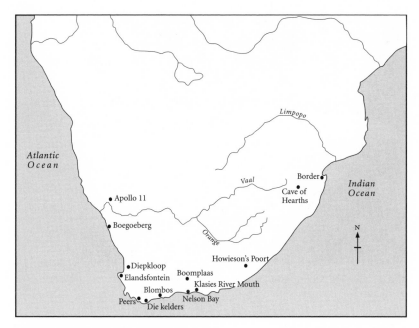

1. Southern Africa, 70,000 years ago

evidently had a greater number and diversity of plant species per square mile than anywhere else on earth—more than 8,000 species, for example, two-thirds of them unique, grow there today—and the oceans were full of mammal, fish, and shellfish species that were unlikely to have been much affected by the drop in temperature. But it would take a newfound ingenuity and cleverness, some serious problem-solving at both an individual and a tribal level, to exploit this diversity enough to survive.

And this is just what the fossil record shows.

Blombos Cave can tell the story of what happened in southern Africa, though it is only one of a dozen caves where evidence has been found. About 30 yards above the blue-green Indian Ocean, some 190 miles east of modern Cape Town, it is a small cave with a long, narrow opening that was occupied for several millennia around 70,000 years ago by people who exemplified the modern culture of the volcanic winter. A team of archeologists led by Christopher Henshilwood of the Iziko Museums of South Africa and the University of Bergen has systematically uncovered since the first dig in 1992 exactly how resourceful and resilient they were.

1. Blombos Cave. Courtesy of University of Bergen.

First, and most important, the Blombos band had a stone-tool technology that was new to humankind: a system of punching off long, thin blades from blocks of fine-grained quartzite or silcrete stone, probably with a wooden awl or punch, and then shaping them with careful stone blows into small, very sharp two-inch points. These would then be fastened on wooden hafts either with some kind of vegetal mastic, traces of which can still be detected on some points, or thongs of hide or vine, which of course degenerate in time and do not fossilize; this is the oldest known evidence from anywhere in the world of composite tools. The resulting spears were weapons sharp enough to penetrate the hide of almost any of the larger local mammals, and could be thrown from a relatively safe distance with enough power to cause serious injury and, if a number of them are thrown at the same time at the right places, death. Along with similar spear points from a dozen other southern African caves in the period around 70,000 years ago, this development marks a true watershed for the Sapiens species.

One indication of the importance of the spear points is that in many places the original stone came from long distances away, suggesting that the artisans knew which stones were best for getting a sharp edge and were willing to invest in the time and trouble to get them. The points of silcrete and fine-

Bifacial points   **Scrapers**

2. Spear points, Blombos Cave, c. 70,000 years ago. Courtesy of
University of Bergen.

grain stones at Blombos, for example, are almost all from non-local sources,
as is much of the ocher; at Border Cave the chalcedony used for nearly half
the spear points came from some twenty-five miles away; at Nelson Bay Cave
on the southernmost coast the proportions of exotic stones range from 83 to
99 per cent over the millennia of occupation there. Such long-distance resource gathering implies a greater degree and wider range of exploitation of
the countryside than previously seen, and even the possibility of a cooperative
trade among different bands so as to add to their repertory of resources (new
foods or flints) and weaponry (new woods for spears). Two anthropologists
who have examined the artifacts of this period, Stanley Ambrose and Karl
Lorenz, have argued in fact that this new stone industry "marks the first time
in human history when there was a significant change in human territorial
organization," another survival response to volcanic winter.

It is possible that Sapiens did some sort of hunting in previous millennia,
but if so the targets would have been small game that might be stunned by
throwing a rock or even caught by casting a net. Meat was an important part of
their diets, but it is generally thought that it was usually obtained by scavenging: taking a half-eaten carcass from a tree where a cat had left it for safekeep-

ing, or chasing away a hyena or a flock of buzzards from some young antelope corpse. And the fossil record suggests that humans normally took back to their camps predominantly the heads and feet of most animals—that is, the parts that were left behind after the carnivores had eaten the choicer ones.

But now the food crisis was acute. New measures had to be taken, new and larger food species sought out, new means of killing developed and perfected. Hunting, and of large animals, now became a fundamental part of Sapiens' survival. The unmistakable evidence of spear points in Blombos and the other caves of southern Africa—which make up tool types known as the Howieson's Poort and Still Bay industries, after two other cave sites—shows that: it is proof, according to Richard Klein, who has worked this part of the continent since 1976, of "a precocious emergence of fully modern human behavior."[3]

But Blombos has even more. In a series of excavations from 1992 to 2000, Henshilwood's team found not just a few isolated artifacts made out of *bone* but evidence of what they call "a bone tool industry," also dating from around 70,000 years ago, the earliest record of the use of this material outside the Katanda site in Congo. Among the tools were three formal, standardized, foot-long spear points, shaped by stone and then highly polished—probably with leather and a stone powder, most likely ocher—and thus aerodynamically efficient and capable of deep penetration into an animal; the other two dozen pieces were smaller points that may have been daggers or awls.

Still more evidence of the primacy of hunting comes from the animal remains at Blombos, which show a great variety of species and a great abundance of specimens. Of the land animals, the most numerous were little ones like the dune mole rat and the rock hyrax (similar to a woodchuck), more than 2,000 samples of which were found, but there were also numerous seals, antelopes from elands to klipspringers, and even evidence of rhinoceros (which would have weighed in at some 3,000 pounds) and hippopotamus (as much as 6,000 pounds). That this was not exceptional is shown by the remains from other southern African caves, which have evidence of wildebeest, hartebeest, zebra, a variety of antelopes, and such decidedly dangerous prey as the giant cape buffalo (which could weigh as much as 2,000 pounds, with a horn span of 6 feet), elephant (9,500 pounds), and bush pig (150 pounds, but fierce). And it is clear from three clues that these animals were hunted and not scavenged: stone cut marks on the bones are abundant and do not overlie the tooth marks of other carnivores, as is typical with scavenged bones; whole carcasses rather than just select bones are present, showing that the entire animal was butchered in the cave; and a large number of bones of young animals (especially buffalo) are present, which are rarely found at scavenger sites because other carnivores typically devour them at the kill.

Of course these early hunters may not have been all that efficient—that is Klein's considered view—and it may have taken several generations to develop the art of effective hunting in the coordinated groups that would be needed to take down a giant buffalo or zebra, but there is no doubt that hunted animals, many of them large and some of them quite dangerous, begin to be an important part of the Sapiens' food source and hence inevitably of their way of life. Judging from the evidence at Klasies River Mouth Cave, the anthropologist Richard Milo has said that these people "were apparently active hunters who produced composite tools and who planned and executed complex tasks within a social framework," in other words, hunters of a "modern" type on a par with, if at first not quite as proficient as, the later Sapiens of Europe. And lest there be any question, the tip of a stone point has been found in that cave embedded in a giant buffalo from this era.

And aquatic species too, in greater variety and abundance than ever before. Some fish and shellfish had been harvested from at least 125,000 years ago—in fact it has been suggested that it was the presence of polyunsaturated fatty acids, Omega 3s and Omega 6s, that was responsible for the dramatic expansion of the Sapiens brain—but now there is evidence, as especially in Blombos, of expanded and sustained seafood exploitation. Seals are the commonest animal—there seem to have been at least 240 individual animals brought to the cave—but there are also dolphins, penguins, possibly whale, shellfish including snails, limpets, and brown mussels, and such large deep-water species, not found in the record before, as red stumpnose, catfish, and black musselcracker (of which there are more than 1,200 bones). In the opinion of the Henshilwood team, a number of the bone artifacts at the site were used for fishing, and the breakage patterns of the bones there indicate that the fish were actively caught rather than scavenged.

Spearpoints. Hunting and fishing. Serious, purposeful exploitation of species, all across the tip of southern Africa. Humankind was beginning that transition into a different kind of being, one who would be an active and purposeful killer of wild animals, even large and sometimes fierce animals, and who was able to think of them as separate beings, distanced, rightful prey that could be stalked and chased and slaughtered and butchered at least often enough to make them a central part of the human diet, and who made the quest for them, in premeditated and powerful groups, a central part of human activity. No longer did these humans see themselves as woven into the web of life, as just another mammalian species, even if erect and clever, as we had done for the preceding two million years, fearing the dangerous beasts and living off their leavings. Hunting changed all that: as the Duke University anthropologist Matt Cartmill put it in his classic study of hunting through history, it "entailed man's estrangement from nature" and created fear and flight in creatures with whom we had lived in rough harmony for eons. Where once we would have moved easily among the other animals, at least the noncarnivorous ones, and drunk at the same waterholes, much as the other species did, now the mostly peaceful relationship was changed. Now we were different, we declared certain other species to be our rightful prey when and how we wanted them, and those species learned to sense danger at the smell of our bodies and flee from our presence. And if we still had to be wary of the other carnivores, now we were their equals in the hunt and occasionally even made them our prey.

The psychological impact of all that, no less than the social, must have been profound. As Cartmill points out, hunting is not "intrinsically pleasurable" for most people: "There is no reason to think that human beings have any innate fondness for bloodshed" and indeed they seem to show "an ingrained *reluctance* to kill," for it "is often entangled with something dark, violent, and irrational in the human psyche." This the Sapiens hunters learned to overcome, pressed by the urgency of survival in a changed world with the volcanic winter and after, but at what cost, with what feelings of remorse, or guilt, or anguish? And by what process of denial or transference or justification did they live, individually and tribally, with those feelings?

It was a gradual transition into hunting, no doubt, and the first phase of it, triggered by the volcanic winter, may have developed over several decades or even more, spreading gradually across the whole bottom of the continent. But when fully adopted and made a fundamental part of daily life, it was to be a profound alteration of the human position in nature, what Barbara Ehrenreich in her study of human violence has called "the single greatest advance in human evolution." If advance it be.

The whole question of hunting in the Stone Age—when it began, who practiced it, how extensive it was—is a complex one, rife with many and sometimes contrary opinions. But I think a few basic truths stand out.

To begin with, it is highly likely that *Homo erectus* for its roughly two million years on earth was primarily a scavenger, though a rather effective and maybe confrontational one at that. Meat and fat were an important part of the hominid diet, particularly as fuel for increasingly larger brains, but various estimates suggest that it may have made up no more than 10 to 20 per cent of total caloric intake, with plants, roots, berries, and nuts making up most of the rest. It was undoubtedly this scavenging, and some control of fire, that enabled Erectus bands to migrate from Africa into colder latitudes: for example into Europe (Dminisi, Georgia, about 40 degrees north) by more than a million years ago and across southern Asia into northern China (Donggutuo, China, also 40 degrees north) a little later. The various stone tools that survive indicate by wear marks and microscopic evidence that they would have been ideal for the processing of meat, and it is probable that by 1.5 million years ago or so Erectus would have had at least occasional use of fire to cook and tenderize meat.

Second, it is unlikely that the Neandertals—which is what Erectus evolved into in Europe and the Levant by around 250,000 years ago—were ever very considerable hunters, at least until they came into contact with Sapiens around 50,000 years ago, and some paleontologists argue flatly that they never went after large mammals like cave bear or woolly mammoth, despite the way they have been depicted in modern reproductions. It is true that preceding the Neandertals there are two early sites indicating some possible hunting in Europe: Schoningen, Germany, a site 400–350,000 years old where eight wooden spears, some only a yard long but two around six and a half feet and one ten feet long, were found a few years back in association with a great many horse bones, though their weight suggests that they were used for thrusting rather than throwing, and Clacton-on-Sea, England, where a foot-long tapered wooden piece was found, dated to the same era, which could have been a short spear or simply served as a digging stick or stake. But these sites are so anomalous that it is almost difficult to take the stakes as weapons at all, and in any case, since nothing similar to them has been found anywhere else before the Neandertals, clearly the practice of hunting, if it existed, was hardly widespread.

As to the Neandertals themselves, the evidence before 50,000 years ago is sketchy. At Lehringen, in northern Germany, a nearly eight-foot wooden stake, again a jabbing rather than a throwing weapon, was found in the ribs of an elephant dating to about 125,000 years ago, and Cotte St. Brelade on the isle of Jersey appears to have been a butchering site around 180,000 years ago, where mammoths and woolly rhinoceroses were dismembered after they were presumably stampeded in what is generally called a "cliff drive," not necessarily by humans, over a nearby headland. A number of sites after 70,000 years ago—Maurens, LaBorde, Combe-Granal, and Genay in France; Wallertheim in Germany; Zwollen in Poland; and Mezmajskaya in southern Russia among them—have bone remains indicating that the inhabitants concentrated almost exclusively on a single species while they lived there, and with a high proportion of prime-age adults, suggesting that there was either confrontational or opportunistic scavenging of huge herds like reindeer and bison or else active hunting, though no hunting tools have been found. And a few sites of the same period and later—St. Brelade and La Quina in France, Staroselje in Ukraine—show evidence (because of the widely ranging ages of the animals) of cliff drives.

Most of the other evidence for Neandertal hunting comes after presumed contact with Sapiens. Several sites in the Levant have likely spearpoints—as at Umm el Tlel, Syria, where Neandertals occasionally used triangular stone points in the period of 60–45,000 years ago—but it seems probable that they learned the techniques from the nearby Sapiens in the region, who may well have learned it from people farther south in Africa if they did not develop it independently. There are unquestioned residues from substances used for fastening stone to wood found at several Neandertal places in southwestern France (55–44,000 years ago), at Buran Kaya III in Ukraine (37–32,000), and at Konigsaue in Germany (around 40,000). And analysis of the diet of Neandertals at the site of Vindija Cave, Croatia, indicates that at least toward the end of their existence and after 15,000 years of living near Sapiens societies (33–32,000 years ago) they were overwhelmingly meat eaters, to an extent that would not have been possible to sustain just by scavenging, and therefore were predators as effective as their Sapiens neighbors.

But whatever the indications are for Neandertal hunting, nothing suggests that it was anywhere at any time the same kind of preoccupation, or special economic and social feature, as it was for Sapiens. Neandertals did not develop the numbers and types of weaponry that the moderns did, or use bone or antler or ivory—in fact they developed no specialized hunting tools at all—nor did they ever go after the wide range of animal and marine types, and except on rare occasions they certainly did not, despite the myth, go after large, dangerous animals like mammoths. They were not hunters with a capital "H" like their successors.

For finally what is clear about those successors is that not only did they become Hunters but they made hunting a centerpiece of their social existence. Hunting, particularly of large and fierce animals, required some sophisticated coordination and cooperation among fairly sizable groups of people, and cohesive social systems would need to have arisen to assure that those groups were well informed, well trained, and successful. This in turn would mean that some ritual, preparation, and education would be necessary to equip young people (presumably young men, as in most surviving hunter-gatherer bands, but possibly young women too, as in a few) for joining the hunting groups—and to learn how to tolerate and justify the act of putting fellow creatures painfully to death. And as meat became central to the Sapiens diet, rituals would no doubt have arisen to promote the success of the hunt and the safety of the hunters, leading to bodily decoration and forms of art. This may have happened only gradually and with different intensities at different times, but by the time Sapiens moved out of Africa after about 50,000 years ago, the array of hunting tools they had developed was so varied and the number of animals killed so plentiful at nearly every site that there could be no question about the primacy of hunting in their lives.

But hunting was not the only reaction to the crisis of volcanic winter. Not surprisingly, this seems to have been the point at which humans started wearing clothes, presumably animal skins and furs. None of it survives in the fossil record, nor is there certain evidence of anything that could be called a needle, though stone dagger or awl points such as those at Blombos could certainly have been used to make holes in skins. The finding is based rather on the thoroughly mundane genetic evidence that the body louse, which is the only louse that lives in clothes, evolved from the head louse (perhaps as a response to volcanic winter) about 70,000 years ago. This additional exploitation of animals is exactly of a piece with the kind of Sapiens thinking evolving in other ways around this time.

An even more dramatic change, according to the theory advanced by Hilary Deacon, an eminent archeologist at the University of Stellenbosch in South Africa, was to use fire to directly change the environment and promote plant growth. The plunge in temperature at the volcanic winter would have hit the surface plants of Africa in a hard way, but it would have had no appreciable effect on tubers and bulbs and corms beneath the earth. At Strathalan B Cave in the foothills of the Drakensberg mountains, a few miles from the coast in the Eastern Cape, two South African researchers have recovered identifiable remains of geophytes, a perennial plant of the region with potato-like bulbs underground that are particularly rich in carbohydrates; Strathalan dates to a later era, around 25,000 years ago, but similar geophyte residues found around the 70,000-year-old mark at the Klasies River Mouth site indicate that the practice was developed much earlier. It is Deacon's belief that people of this age came to depend on these geophytes for much of their food, and that they would have known, presumably in the aftermath of a natural fire, that these bulbs proliferate after the surface vegetation is burned off. (Modern research indicates that the plant is eight times more numerous after firing.) When much of their regular plant diet was no longer available, the assumption is, they learned to concentrate on the geophytes and figured out how to set fires annually to force new growth, as many African peoples do today.

This "farming with fire," Deacon thinks, was "a key change" in the behavior of modern Sapiens. But "key" seems a bit inadequate to describe this: for this seems to be, as far as we know, the first time in prehistory that humans had the will and means to consciously manipulate their habitat in a concerted way, to establish some deliberate control over a major food source much as agriculture would enable them to do many millennia later. The psychological effect of this, quite apart from the nutritional, would most likely have been immense, conferring on these people a sense of power over their environment that when coupled with their newfound skills at hunting and fishing, must eventually have led them to a new perception of their uniqueness in the world. Such new perception is suggested, with clear indications of true modern behavior, by several other elements of the fossil record in these southern African caves.

Perhaps the most intriguing, thanks once again to a find at Blombos Cave, are two small pieces, two inches and three inches long, of slate-like red ocher with a set of crossed X's scored by three horizontal lines, dated to about 70,000 years ago. Whether or not this is "art," as some archeologists maintain, it is certainly some kind of purposeful abstract form, possibly even a record of some kind and the earliest such representation known unambiguously from human hands.[4] (A deliberately notched ocher piece has been found at the Hollow Rock Shelter, but its dating, though in this same time span, is less precise.) Christopher Henshilwood, part of the team at a later Blombos dig that reported on these pieces in 2002, believes that together with small etched marks on the bone spear points that his team found, they are evidence of "symbolic thinking," because there is clearly "a system" to the engraved patterns: "We don't know what they mean," he says, "but they are symbols that I think could have been interpreted by those people as having meaning that would be understood by others." And such "communicating with symbols," as Richard Klein has noted, "provides an unambiguous signature of our modernity—once symbols appear in the archeological record . . . we know we are dealing with people like us."

If these engraved pieces are evidence of symbolic thinking, and communication, and not just idle doodling, they represent a very significant milestone for the human species. I think it probable that earlier Erectus peoples had an elementary knowledge of symbols—a certain footprint would indicate,

3. Incised ocher, Blombos Cave, c. 70,000 years ago. Courtesy of University of Bergen.

and stand for, a certain animal—but what paleontologists mean by symbolic thinking is a more complex reordering of nature in the mind to include a regular division and classification of its elements, with a sense of the world beyond the individual and the group, and a sense of time and place beyond the here and now. It indicates that people were able to make a separation of ideas and objects, representation and reality, in what could have been a completely new, and what we would recognize as a modern, way. That, in turn, would allow the evolution of concepts of Self and Other that lie, for example, at the heart of the whole hunting enterprise and may even be the essential distancing that permits regular and frequent killing of animals: that over there is an eland, it is an *animal*, and I am a hunter here who will eat its flesh. (It was also the origin, according to Rousseau, no paleontologist, of evil in the world.) And it seems probable that this symbolic ability and the kind of social communication (and cohesion) that it suggests were an essential reason for the success of the southern Africans in the face of the stress that these societies endured in the volcanic winter crisis and after.

Communication by symbols of course suggests language, and here we come to a thicket of nettlesome complexity that has engaged paleontologists for decades, most acutely in the last few years. The problem is that language, as they say, does not fossilize, and not until there is obvious writing is there any sure evidence for its existence. But it is possible to make inferences from artifacts and the behaviors that they indicate about humans' ability to communicate, and any number of schools of thought have grown up around the poles of different inference makers. It is not necessary here to plunge headlong into this thicket, or to worry about who first had language and when, but the clear indication of symbolic thought indicates that we can reasonably surmise some kind of developed language in use around 70,000 years ago.

Humans probably were anatomically able to make sounds as early as the first Erectus, as indicated by the brain configuration, particularly a section associated with speech called Broca's area, and a likely position of the larynx low enough to permit some degree of articulate speech. By the time modern humans emerged, the body had a full capacity for speech in both brain configuration and upper respiratory tract, and the ability for rapidly spoken phonemic speech such as we know today, but when humans actually produced language is not known. Some have suggested that speech did not evolve until 45,000 years ago or so, when modern culture is found in both Africa and Europe, but the increasing evidence for an earlier modern culture origin that I have suggested here, especially the evidence of symbolic thinking, allows the possibility of language much earlier. The prominent English archeologist Paul Mellars has argued that "social complexity," including increased hunting, long-distance trading, and the standardization of distinct tools such as the Howieson's Poort industry, is prime evidence of language. On much the same evidence Ian Tattersall of the American Museum of Natural History has suggested that "if a cultural innovation occurred in one human population" in Africa around 70–60,000 years ago—and as we have seen, it did—this would have "activated a potential for symbolic cognitive processes that had resided in the human brain all along" and that then spread by cultural diffusion elsewhere. And one interesting piece of support for this is the recent detection of a "language gene" (FOXP2), conferring the ability of rapid articulation, that can be traced back genetically to an original mutation during the last 100,000 years or so. If Sapiens did indeed have language, that would do a great deal to explain their success at survival and eventual conquest of the continents.

Found along with the engraved pieces at Blombos was a fragment of mammal bone that likewise had been intentionally carved with a stone tool, and more than 8,500 pieces of ocher that the Henshilwood team takes as additional indications of art, both because ocher is often associated with art and decoration later in prehistory and because most of these pieces show evidence of having been scraped to make a powder, perhaps for pigments. There are tantalizing earlier indications that humans used ocher and a similar hematite for decorations—ocher fragments from the Kapthurin Formation in Kenya, for example, have been dated to as early as 280,000 years ago—but the abundance of the material at Blombos represents a quantitative change. More than a hundred other ocher fragments, many tapered to resemble pencils, were found at the Klasies River Mouth site at the Howieson's Poort levels, and grindstones with ocher traces that might have been "palettes" have been found at Die Kelders Cave.

Also at Blombos the Henshilwood team in 2004 made still another remarkable find: what is thought to be the earliest evidence of personal ornamentation, forty-one drilled and polished pea-sized beads made from the shell of a small, snail-like mollusk, found in clusters that suggest they were strung together as necklaces or bracelets or perhaps sewn on clothes. (There were also twenty-one pieces of ocher that were perforated in neat circles, probably by marine organisms rather than human drilling, but that could well have been used as jewelry nonetheless.) Such an innovative expression argues that the stress of the new climatic conditions inspired another survival response, this time on the social or ritual level rather than the material or alimentary level. Decoration—of either bodies or clothes—implies first of all some kind of self-recognition and, more than that, personal *self-assertion* within a tribal band, a statement of ego that may have been born of the trauma of volcanic winter (psychologically, narcissism is a common result of trauma) and a new need for individualized responses of a sort that would have been completely foreign to *Homo erectus*. It may also reflect a need for the first time for bands and tribes to distinguish themselves from others, a means of identity and identification for peoples with the increasingly wider ranges of hunting and trading brought on by the pressures of survival in difficult times.

Altogether this might not seem to be abundant evidence for art, decoration, or symbolic representation, at or near the glacial winter, but what there is in the record is clear enough and marks a decisive change from earlier eras. If it seems annoyingly meager, part of the problem is simply that Africa as a whole has only started to be comprehensively surveyed and professionally excavated in the last few decades, and even with several dozen sites now being explored the numbers pale compared to those of Europe.[5] It is possible also that at this point in prehistory the number of people, and therefore the number of occupied sites, was not so great, a population die-down that may have been a consequence of the severe climate and loss of vegetation. Then too, evidence indicates that many places were abandoned somewhere around 60,000 years ago, meaning that many of them would yield indications of only a few thousand years of human impact.

Nonetheless there seems to be enough already in the record, taking one thing with another — hunting, fishing, long-distance travel, clothing, hearths, fire management, symbolic thinking, art, decoration, social cohesion — to argue forcefully for the evolution of a new kind of human culture, recognizably modern, at about the critical 70,000-year-ago juncture, at least in southern Africa as far north as Zambia. The people of this culture had abilities and mental facilities very much like those found later in Europe and Asia, and with the possible exception of burials, for which the evidence in Africa at this time is very weak, they displayed much the same kind of behavior. As Hilary Deacon has summarized it, they "had essentially the same perception of their environment" as people 60,000 years later, had the same "subsistence behaviour" in relying on meat, fish, and tubers, had the same "ability to solve problems relating to resources," and had a "use of artifacts as symbols to cope with stress that indicates a modern quality of behaviour."

What happened to this modern culture is not known well at all because the number of excavated sites between about 60,000 and 40,000 years ago in Africa is small. All that is known for sure is that the southern African stone weapon industry (the Howieson's Poort industry) seems to last for about 10,000 years and then vanishes, and that the evidence for human occupation in southern African sites declines sharply after 60,000 years ago, as if most of the area was abandoned.

In the shaping of a hunting way of life there were side effects that served to make human survival and longevity more likely than ever before. The inclusion of polyunsaturated fatty acids from seafood improved both fetal and maternal nutrition considerably, assuring healthier infancy, and the overall diet, markedly increased with many new plant and animal foods, provided more resources for growing children. This new expansion of the diet would also have meant a decline in general mortality, with fewer periods of scarcity and starvation, while a longer life (men to fifty or sixty, women to forty or fifty) would have allowed grandparents to care for children, enabling mothers to return to productive work earlier. Long-distance connections to other bands and tribes might well have provided a resource cushion in times of local disruptions and depletions. Hunting itself was relatively safe, except for the fiercer animals like the buffalo and bush pig, with spears that could be hurled from a distance and possibly arrows (though there is no trace of bows in the fossil record until around 12,000 years ago) propelled from even farther away. And the kind of complex and developed social life suggested by decoration and symbolic activity may have provided the kind of cohesive and cooperative atmosphere that would enhance a band's survival.

All in all, there is every reason to concur with the assessment of the anthropologists Sally McBrearty and Alison Brooks that there was "dramatic population growth in Africa" around this time. (The archeological evidence is supported by a theory, based on the probable evolution of organic compounds called nucleotides, that a Sapiens population expansion occurred about 65,000–35,000 years ago.) This, coupled with the extensive exploitation of their environments allowed by improved hunting techniques and the use of fire, might well over time have exhausted the Sapiens' local resources, both animal and vegetable, and compelled them to move off in search of new territories. It was their very success, in other words, that led to the abandonment of their homes, as would happen over and over again in the succeeding years of a species that never seemed to learn how to live within the limits of its given ecosystems.

In addition, as the Greenland ice-core evidence for climate change indicates, there was a period of increasing warmth beginning rather suddenly around 57,000 years ago, and lasting, with colder intervals, for approximately the next 20,000 years. That would have led to the migration of surviving game herds into other habitats, as happened in previous warmer and wetter periods, forcing human populations to follow and invade new territories in central and eastern Africa in search of new niches to settle.

It is well-nigh impossible to trace these migrations, for the archeological record at this period is so weak. At best we can find a few sites elsewhere in Africa, and then in Asia and the Levant and Europe, where other examples of modern culture turn up in the years after 60,000 years ago. There is usually no way of knowing if these artifacts were brought from the South over these millennia or if they were the independent creations of other Sapiens, in response to either their own experience of volcanic winter or other climatic or social crises. But that they do not appear, with a few exceptions, until 10,000 years or more after they emerged in southern Africa allows the hypothesis at any rate of cultural migrations northward.

East Africa, which may have been continuously settled from 250,000 years ago onward, starts to show signs of modern culture around 50,000 years ago. Mumba Shelter, near Lake Victoria in modern Tanzania, for example, has some modern-looking stone tools dated to between 65,000 and 45,000 years ago, as well as ostrich shell beads, unambiguous signs of decoration, that give dates of 52,000 years and 45–40,000 years ago. Thirteen other ostrich eggshell beads, about a quarter of an inch in diameter, have been found at a rock shelter known as Enkapune Ya Muto ("Twilight Cave") in the Rift Valley of Kenya and carbon-dated to about 40,000 years ago but thought by their excavator, the archeologist Stanley Ambrose, to be about 45,000 years old. Although these ostrich-shell beads are flat, they are similar in size to the Blombos Cave mollusk-shell beads and were also probably strung together to make necklaces or bracelets. They provide suggestive support for the idea that the people moving from southern Africa may have taken their practices of decoration with them as they moved to the new and fresher lands to the northeast.

It also appears that modern influences, and possibly modern people as well, moved into northern Africa at about this period. There, sometime around 60–50,000 years ago (consensus but not absolutely sure dates), a new stone-tool industry and hunting culture appears in a swath that runs from the Nile Valley into the Sahara and across the continent to Mali and Mauritania and, with the largest number of cave sites, along the Mediterranean Maghreb in the northwest from Morocco to Tunisia. It has some elements of earlier industries but is particularly marked by projectile points that have a small shaft, or stem, at their base—giving the projectile the shape of the familiar American Indian arrowhead—that was used to make attachment to a shaft easier and more durable. (One anthropologist has also suggested that this culture also invented a spear thrower around 40,000 years ago, but that is still unproven.) This culture—known technically as the Aterian after a site, Bir el Ater, in eastern Algeria—was successful for tens of thousands of years, until the onset of colder, drier climates around 30,000 years ago (and in a few places like the Maghreb until 20,000 years or so), attesting to the remarkable ability of Sapiens culture to adapt to an extraordinary range of territories. Aterian sites are found in deserts (at oases and streams, to be sure), in rough, craggy mountains, along the Atlantic ocean shore, in the savannas along the Mediterranean, and next to rivers large and small—indeed there is no ecosystem of northern Africa which these people did not succeed in exploiting.

It has been thought for some time that this culture—its influence, not necessarily its people—moved up from central Africa after the volcanic winter and its aftermath, when a warmer, wetter climate made parts of the Sahara and the Mediterranean littoral more habitable. The evidence is not abundant, but it is generally thought that the route of influence for stone toolwork—spreading the art and craft of spearpoints, for example—went in the West from Chad and Niger up to the Libyan coast and hence to the Moroccan Maghreb, and in the East up the Nile valley into the Levant and then westward along the Mediterranean. As one early investigator analyzed it, "the origin of the classical Saharan and Maghrebian Aterian should be sought in Central Africa." This might suggest that we could look to the southern African peoples who we presume moved northward 60,000 years ago as the distant source of the Aterian stoneworking, but the fossil record for central Africa is so poor that it is not possible to confirm this. There is an identifiable stone industry in the Congo basin called the Lupemban, with quite large spear points, but it is poorly dated and one can only assume that it shows the influence of southern cultures because, for example, the points look in particular like those of the Still Bay site in South Africa. But they are not stemmed, so that feature presumably evolved in the course of the dispersals northward.

The phenomenon of extensive migrations of people over many millennia, presumably in successive waves of fairly small bands, should not seem surprising or anomalous. In the first place, we know from historic times that any number of external forces induce small populations to move: population growth, exhaustion of lands, years of limited food supplies, flight from diseases, pressures from neighboring peoples. And second, the record shows that various humans not only moved through Africa but also moved *out* of Africa at a number of times in prehistory.

Early *Homo erectus* bands expanded out of East Africa onto the Ethiopian plateau at least 1.5 million years ago and went into the Levant ('Ubeidiya, Israel) by about 1.4 million years and even into Europe as far north as the Caucases (Dmanisi, Georgia) by about 1.8 million years. Other Erectus populations migrated into Asia, perhaps traveling eastward out of the Levant to the Tigris and down along the fertile shorelines into India and beyond, or possibly across the Red Sea at the Bab el Mandeb strait (which would have been very narrow at the coldest times, when glaciers locked up large amounts of seawater and sea levels were low) and then along the coasts to the subcontinent. In any case, Erectus people reached as far east as the Indonesian islands (Mata Menge on Flores, Sangiran on Java) and up into northern China (Nihewan Basin and Lantian) at least by 1.6 million years ago. Later peoples moved into Western Europe by at least 800,000 years ago (Atapuerca, Spain) — probably migrating through the Levant, though possibly across the Gibraltar strait in times of lowered sea levels — and by 500,000 years ago had settled across western Europe, from southern England (Boxgrove) to northern Italy (Visogliano).

And then, sometime after 60,000 years ago, it was the Sapiens' turn.

One stream migrated into Asia, most probably across the Bab el Mandeb passage and along the coastlines as their Erectus forebears 800,000 years earlier had done (very few archeological sites along the coasts actually confirm this supposition, but most sites would now be submerged 30 or 40 yards beneath the ocean), and reached the northern coast of Borneo island (Niah Cave, where a Sapiens skull has been found) by about 45,000 years ago, and northeastern New Guinea (Bobongara, where hafted axeheads have been found) around 40,000. The evidence is clear, according to the prehistorian Rhys Jones of the Australian National University, that "a fundamental change in the archaeological record . . . occurred in Southeast Asia" about 40,000 years ago, reflecting "changes in human behaviour on a global scale."

It has been argued in some recent studies of migration movements that "the time required to cover the southern coasts of the Asiatic continent can be estimated at approximately 10,000 to 15,000 years," though the dispersal would have gone in successive waves as one area was overpopulated or overexploited after many years and then some bands were forced to move on. That would confirm the idea that the migrations from Africa began after 60,000 years ago, perhaps sometime around 55–50,000 years. And that possibility gets further weight from several recent genetic studies proving that modern Asian populations can be traced back to African populations and that migration from Africa, initially into Asia, began fairly recently, in archeological terms. One study in China, for example, of the Y chromosome in 12,127 Asian men determined that it came from an African source, between 89,000 and 35,000 years ago (a mean of 62,000), another of people in India and the south Arabian peninsula suggested that the exodus from Africa was "more than 50,000 years ago," and a third in Sweden of the DNA of 53 individuals from around the world confirmed an African origin for Sapiens and found that "the migration from the continent began about 52,000 years ago."

There is, however, one hitch: Australia. The earliest absolutely confirmed dates for modern human presence in Australia start at around 48,000 years ago (Devil's Lair) and 46,000 (Lake Mungo), with others somewhat later (Carpenters Gap, 39,000 years; Upper Swan, 38,000 years), which fits well with the picture so far (assuming that a migration time of about 7,000 years is plausible). But two recent dating measurements at rock-shelter sites on the northern coast (Malakunanja II and Nauwalabila I) imply a human antiquity between 60,000 and 50,000 years ago (along with large pieces of red and yellow ocher), which does not fit as well. It is of course possible that the wave of migration left Africa earlier than 55,000 years ago and made its way to Australia rapidly, leaving little trace of itself in between; two recent studies of mitochondrial DNA in two "relic" Asian populations suggest that their ancestors might have arrived there between 63,000 and 44,000 years ago (a mean of 53,500 years). It is also possible that the earlier rock-shelter settlements were not by Sapiens at all but by remnant Erectus people, who did survive in other parts of Asia until about 30,000 years ago, and that they had the capacity to cross open water from the Indonesian islands and used ocher for either decoration or, more likely, tanning hides. And there is some reason to question the 60–50,000-year date for the two rock shelters, since the dating methods depend on measuring energy in quartz grains found with artifacts, and there is a real possibility that these grains have become displaced over time and actually well predate the tools; no archeological site in Australia has given a radiocarbon date of more than 40,000 years ago.

One piece of evidence for the idea that a modern hunting culture did not reach Australia until 48–46,000 years ago is an intriguing one, laden with implications that we will consider later. It is the indication that there was a wave of large-animal extinctions after 46,000 years ago, eliminating nineteen genera and some sixty species—thought to be 86.4 per cent of the large animals then alive on the continent—in a fairly short time, perhaps one or two thousand years. The effects of climate change have been suggested as one possible cause, but the fluctuations in the period from 46,000 to 25,000 years ago were not particularly severe and not markedly different from those of the preceding 30,000 years. The one effect that would be decidedly different was the arrival of modern Sapiens with their need for, and skill at, hunting, and probably their management of fire: the first would have made them devastating predators of large herbivores with no previous experience of humankind, while the second would have made them severe disrupters of long-stable habitats and food sources up and down the food chain. It is easy to see that the large marsupials like giant kangaroos, flightless birds like giant emus, and slow-moving tortoises the size of a small car would be attractive and easily huntable prey, particularly since humans by then had had 25,000 years of perfecting their weaponry, and it is not hard to imagine that small populations would dwindle beyond the point of recovery very quickly. The loss of many herbivore species would have a terrible effect on the carnivore species that depended on them—two kinds of large cats, for example, and a giant variant lizard—and these too would have been driven to extinction. And if fires were set either to encourage vegetation growth or to drive animal herds, as done by the people of nearby New Guinea at the forest site of Kosipe 30,000 years ago, the destruction of plant cover and forage for the herbivore populations, especially the giant animals that needed extensive ranges and abundant supplies, would have been equally ruinous.[6]

Humans always have an effect on their environment, but the modern Sapiens, shrouded in a culture that celebrated the hunt and permitted a separation of self from nature, have made a far more severe impact wherever they have gone. Earlier humans, as Niles Eldredge of the American Museum of Natural History has put it, were "very much integrated into their ecosystems." But when Sapiens "began to spread around the globe . . . there really was a difference [and] the world's ecoystems began to feel it; we began to have a direct impact on nature." Direct, yes, and devastating.

And it doesn't end in Australia.

TWO

The Conquest of Europe

55,000–20,000 YEARS AGO

The other long dispersal of modern Sapiens out of Af-rica on its journey to dominate the earth was northward, to the Levant and Europe and central Asia. They were most likely people from East Africa who migrated along the Nile corridor through modern Sudan and Egypt, but they may also have been a branch of the wave of Asian migrants who went up the narrow fertile western corridor of the Arabian Peninsula after the Bab el Mandeb crossing.

Imagine a Sapiens group—a tribe of five hundred, say, in bands of twenty-five or so—living around 55,000 years ago in the lowlands near the headwaters of the White Nile in what is today southern Sudan. They are the inheritors of the modern culture that has spread from southern Africa, and they survive with the skillful hunting and fishing techniques developed over the millennia, the close-knit social organizations that establish and maintain group harmony, the communications capabilities of at least a rudimentary language, and a healthy diet based on both plants and animals in abundance. But there are other inheritor bands around, for the region is fertile and the climate generally benign, and they continue to grow in population—and this means that in time it gets harder and harder to find new fields of tubers, or large herds of impala, or the usual swamp tortoises. Human pressure on the area is pushing it past its carrying capacity, and relations with other bands in other tribes become increasingly stressful as competition intensifies. The tribe decides that it is time to move on, down the river another fifty miles or so, to new territory and fresh resources.

This is how we may assume that a number of Sapiens populations, under the pressure of rapidly increasing growth rates and overexploitation, slowly migrated out of eastern Africa in the millennia after about 55,000 years ago and moved in successive waves down the Nile until they reached the Mediterranean. Studies of prehistoric migrations in Australia and North America and computer simulations of African migrations suggest that what is called a "hiving-off" rate of 25–35 miles every 20 years or so was common, and at that rate Sapiens could have gone the whole length of the Nile in no more than 1,600 years. Even under harsh climate conditions the river and its banks would generally have offered adequate food and, as the anthropologist Nicolas Rolland has noted, the Nile Valley "must have provided a perennial corridor and an attractive natural habitat in which different human groups would converge."

And this is what the archeological record seems to show, allowing for the fact that it is not particularly abundant because most settlements were surface sites rather than cave sites, and thus prone to wind erosion and subsequent human disturbance over the years in a way that caves are not. The Belgian archeologist Philip Van Peer, part of a team that has spent nearly three decades of research there, has argued convincingly that what he calls a "Nubian culture" with "complex cultural features" was established in the region well before 40,000 years ago by "new groups coming from the south" with many of the characteristics we have seen in the southern African culture.

From such sites as Nazlet Safaha and Taramsa 1 nearby, on a bend in the Nile some 300 miles south of present-day Cairo, Van Peer has found evidence confirming "the importance of hunting in subsistence strategies of these groups," indeed "the major subsistence strategy." He bases his conclusion on the extraordinary number of projectile points there—he estimates that people at the two sites dug out 500,000 stone cores from the surrounding flint deposits for knapping into points—and the high proportion of points as compared to other tools, making up 90 per cent in the deposits at Taramsa 1. The number of points, and the extensive quarries from which the stones were taken, also suggest a considerable population density and, to Van Peer, "a context of enhanced social interaction" such as characterizes complex Sapiens cultures elsewhere. A developed modern culture is also suggested by the "lithic complexity" of the stone tool kit and by the way that its artisans would plan ahead to take stones to particular places and knap them for particular purposes.

A single Sapiens skeleton has been found at Taramsa, tentatively dated to 55,000 years ago but possibly as young as 50,000. Besides giving proof that Sapiens were indeed responsible for the Nubian culture, it may represent the earliest instance of a modern human burial. The skeleton, of a young girl of ten or so, was found in a sitting position, facing east, with her head bent back so that her face was skyward, and this unusual posture convinced Pierre Vermeersch, another Belgian scientist who led the team that discovered her in 1998, that she was given a deliberate burial. If so, this would indicate a culture with some idea of an afterlife and therefore presumably something we could call religion, a cultural marker that does not occur in the fossil record before this; although several dozen Neandertal *interments* have been uncovered, none have any signs of being more than practical disposals to get corpses out of the way and prevent them from attracting scavengers, and hence they do not rise to the level of ritualistic *burials*. It is the presence of what anthropologists call "grave goods"—ornaments, artworks, tools—accompanying the bones that is generally taken to mark a true grave; the absence at this Taramsa site of such goods and of any other artifactual evidence for religious ritual, and the fact that this is a lone example 25,000 years before certain burials appear in the record, incline me to doubt that this disposal was a true burial.[1]

We cannot be absolutely sure that it was these Nubian people who gradually expanded and moved into the Levant, but it would be a journey of only six hundred miles or so from the Nile fossil sites, and the migration would match the pattern of earlier African hominid migrations, including the one that led to Sapiens settlement there in the period 110,000–70,000 years ago when Africa was especially hot and dry and many African mammals were drawn to the moister Levant. At any rate new populations moved there sometime not long after 50,000 years ago, and judging by the kinds of toolkits found—typically marked by spear points (some with Y-shaped patterns like those in the Nile) and other sophisticated tools—those people came from Africa. Richard Klein of Stanford takes these tools to be a logical indication of what we know of the Sapiens path: "An African origin," he says, "is . . . implied by the appearance of broadly similar modern behavioral markers in southwestern Asia in the interval between their earliest appearance in Africa and Europe."

Support for this view comes additionally from the kinds of skeletons found, which are typically those of tropical-bred humans, with long limbs and a tall torso providing large surface areas of skin with which to lose heat, unlike those of people bred in cold weather like the Neandertals, short and stocky. And it would seem to be clinched by recent genetic evidence comparing markers on Y chromosomes showing that one of them in northeastern African populations is identical with one in most Levantine people, and that the mutation giving rise to it can be dated to some time around 50–45,000 years ago, meaning a northward migration at about that time.[2]

The first sure evidence of a modern culture in the Levant, from about 49,000 years ago,[3] is the stone tools found at a site called Boker Tachtit on an ancient river plain in the Negev Desert at the southern tip of modern Israel. The objects there include a great many triangular spear points, though for some reason not as precise as the southern African points, and several flint tools that will mark the modern culture as it develops in the Levant and Europe: *blades* (defined as stone flakes at least twice as long—typically about 2 inches—as wide), some of them purposefully blunted on one side so as to make them easier to hold, *scrapers* with one side hammered to a flat, knifelike edge, and pointed, hand-held *burins* used as perforators. And as time went on at this site the tool making rather rapidly became more and more sophisticated, until by 38,000 years ago, when the site was effectively abandoned, the artisans could make all their tool types from a single pyramidal core piece, meaning a great deal more cutting edge from a given stone core than in any earlier era, a technological feat that likewise becomes a hallmark of modern culture.

This expanded and remarkably standardized toolkit—the predominant spear points of several different types, plus blades, bladelets, sidescrapers, endscrapers, burins, toothed crescents, awls, punches, mortars, spatulas, and grinding stones—is a sign of the intensity with which the Boker Tachtit people were able to exploit their surroundings. They were hunting a variety of animals large and small, cutting carcasses easily, scraping animal hides for clothing and maybe shelter, piercing leather to make sewing holes, scraping and cutting wood and vines, grinding plants and tubers, and in general developing instruments to carry out any task of domination they conceived of. And it is also a sign that technology itself was coming to play an essential part in the culture in a markedly new way and that Sapiens society was becoming dependent on (and proficient with) technology to a degree not seen before in hominid history.

One scientist, the computer expert Sheldon Klein at the University of Wisconsin, has even pointed to these tools from Boker Tachtit to argue that Sapiens were developing "a new pattern of cognition," the key to their subsequent success and territorial expansion. He sees the way the knappers went about preparing and then striking their stone cores to create a wide assortment of tools as a special form of reasoning and thinking ahead, particularly the way they could transform the cores for a great variety of other uses with a few deft strokes whenever the need arose. This "analogical reasoning," Klein suggests, in which one thing could be made to stand for many others, was part of the Sapiens' language structures and indeed their way of ordering the world around them and establishing their complex social organizations. Some elementary form of this kind of thinking may well have existed earlier, judging by the complex but much smaller toolkits from the Howieson's Poort technology on down, but it rises to a height at about this time and will go on to characterize the rapid development of increasingly complex and versatile tools that emerge in the fossil record with the sweep of Sapiens from the Levant into Europe.

There is, alas, no certain evidence again of who it was who made these tools at Boker Tachtit, for the first Sapiens bones in this area are dated to about 46,000 years ago—a child's remains at the Ksar 'Akil rock shelter near the coast in what is now Lebanon—though it is assumed that they were not made by the Neandertals who had been in the region but seem to have died out there 50,000 years ago and whose tools, other than the suggestion of some projectile points after about 56,000 years ago, are not of the modern Sapiens kind. And since the tools found at the Ksar 'Akil site were like those of Boker Tachtit, logic suggests that it was Sapiens who made those initial tools 3,000 years earlier.

The Ksar 'Akil skeleton, incidentally—for some reason named "Egbert" in the literature—is another held by many to be the earliest example of a conscious Sapiens burial, the first in the Levant. Burial is implied because the skeleton was overlain by a pile of cobbles deliberately brought into the rock shelter, suggesting some form of gravestone, but it could also have simply been a means of protecting the body from scavenging by local carnivores and have no ritual connotations whatsoever, and again there are no ritualistic grave goods. In short, the Levant, like the Nile Valley, shows no absolutely convincing evidence of burial until much later.

Other sites in the Levant confirm, with exclamation points, the existence of a modern Sapiens culture in the millennia after 49,000 years ago.

Some sort of unquestionable art, for example, emerges at about 45,000 years: a limestone slab and a stone, both smeared with red ocher, found at Qafzeh Cave near the Mediterranean coast of modern Israel dated to 44–42,000 years; small limestone pieces with red and black paint at the nearby Hayonim Cave Level D dated to 32,000 years; and at the same level two small engraved slabs. One of them is arguably the first (and a very rare) depiction of an animal in the Levantine record (and at about the same time as animals appear in European art): a horse in outline with a small nick for an eye, overlain with a series of slashing straight lines that could well represent, as the investigator Alexander Marshack thinks, symbolic "killing" by darts and spears. Since the depiction was covered with red ocher in the middle, possibly representing blood, it clearly had some greater significance than simply decoration, or "art" for public view, and it is hard not to think that it must have been used as an instrument of what anthropologists call "hunting magic," periodically incised before an expedition for luck or power in the hunt. Not all art in

4. Limestone slab with engraved horse, 29–27,000 years ago. Photo copyright the Israel Museum, Jerusalem, by P. Lanyi.

the Sapiens record is necessarily used for hunting magic, as we will see later, but in this case the association seems clear: an example of the relationship of human to animal both utilitarian and antagonistic that is at the heart of the hunting culture.

This emergence for the first time of magic in Sapiens culture—the term "magic" is clumsy, but it can stand for the use of symbolization in specialized tasks designed to exert human control over others—reveals two crucial elements of the Sapiens worldview at this juncture, both central to the process of the domination of nature, and especially important in that both in some ways still underlie our contemporary perspective.

The first was an emphasis on the attempt, as Freud put it in his examination of magic in *Totem and Taboo*, to "subject the processes of nature to the will of man"—not just the hunted animal, as in this case, but presumably the seasons, weather, rivers, and the like, as with historical tribes—by manipulating representations of natural phenomena, with the idea of influencing the real-life counterparts for individual or group benefit. You can do this only when you have a deep-seated conviction that humans have not only the *power* effectively to intervene in nature—to use a slab of stone to represent the killing of a horse gives you the ability to then go out and kill one—but the *legitimacy* of doing so, to achieve what J. G. Frazer in his classic work on magic called "a sovereignty over nature." Freud remarked that this principle, standing behind all magic, was "the omnipotence of thought," the notion that the world was governed not by independent physical laws but by human mental constructs, and he noted that this delusional view in contemporary terms would be evidence of neurosis.

The second element of magic was a belief, not necessarily always conscious, in some kind of supernatural, or nonmaterial, powers, the ones that could be called upon to carry out the wishes that humans expressed through magic ritual. *Supernatural*: it is no longer the natural world that is the source of fulfillment, of guidance and direction, of numinous inspiration, but someplace else, in the imagination, a human construct apart from the grounded world. Calling upon the supernatural suggests beliefs that come very close to religion here, even if not yet necessarily involving gods, and though there are no other direct indications of religion in the Levantine sites of this era, this one small slab suggests that the culture is moving in that otherworldly direction.

Ornamentation also emerges early in the Levantine culture, and copiously too. At Üçağizli Cave, a coastal site on the Mediterranean in a region of southernmost Turkey still ecologically in the Levant, a layer that has been dated to 44–41,000 years ago (but may well be even older) has yielded nearly sixty perforated marine shells used for beads and pendants; subsequent occupiers of the site continued this tradition for thousands of years, and by 35–32,000 years nearly nine hundred examples survive. The beads are quite small (no bigger than three-quarters of an inch), selected for luminous or bright colors or for striking patterns, perforated with a pointed tool, and presumably strung with a vine or cord or conceivably sewn onto clothing; there is also a single example of a claw bone of an eagle or vulture perforated to hang as a pendant.

Another site, the Ksar 'Akil rock shelter some one hundred miles south, has also provided an abundant trove of similar shell beads that are at least as old (possibly even 47,000 years old, though the radiocarbon dating technique is unreliable here), indicating that the culture was fairly widespread and successful; after 32,000 years ago, according to the Princeton archeologist Peter Bogucki, "the acquisition of marine shells for personal ornamentation was

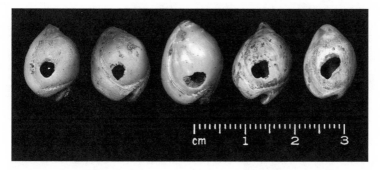

5. Marine shell beads, Üçağizli Cave, 44–41,000 years ago. Courtesy of Mary L. Stiner.

particularly widespread in the Levant." Beads of other materials—deer, bear, and horse teeth, for example—are also found at later sites (Hayonim Cave and El Wad in northern Israel, erq el-Ahmar in Jordan), and at 27,000 years ago, at Ksar 'Akil again, regular incisions on a bone point (perhaps an awl) indicate that ornamentation was extended even to tools.[4]

The lead anthropologists at the Üçağizli site, Steven Kuhn and Mary Stiner of the University of Arizona, draw further conclusions from what they call the "ubiquitous" beadwork, similar to those suggested by the ornaments we have seen from southern and east Africa. For one thing, it implies "a shared system of communication," signaling to people in other bands and tribes something about the wearer's status—group identity, position, wealth, bonding relationship, hunting prowess, or some such—that others would be assumed to understand: "We might expect ornament technology to arise first where the chances of meeting strangers, and the benefits of advertising one's identity and status from afar, were relatively high." It is also a means of distinguishing one band, one tribe, from another, perhaps proclaiming *territoriality* over a hunting range, in the increasingly crowded regions with favorable climate, and a way of proclaiming who you are and where you belong—and who is protecting you.

Such large populations in this region suggest to Kuhn and Stiner that there must have been a significant pressure on humans to "enlarge their dietary repertoire" by increasingly hunting different kinds of animals, and in greater quantities. In their research at Hayonim Cave in northern Israel they determined that at first such slowly reproducing and easy-to-catch species as tortoises and lizards (along with easy-to-scavenge ostrich eggs) made up as much as 52 per cent of the specimens at some levels. But the absence of their remains from overlying levels suggests that they were over-hunted and eventually eliminated from the region, creating "chronic shortages" for many millennia after about 44,000 years ago — another example of a Sapiens inability to understand limits in its capacity for exploitation.

In response to these food shortages some Sapiens bands migrated out of the area, heading north, but the record shows that many chose to stay and just hunt fleeter and faster-reproducing animals like rabbits, partridges, and birds, even though that meant a great deal more time and effort. (That also almost certainly meant the invention of traps and fowling gear, probably nets woven of grasses or vines, though because they do not fossilize they are not evident in the record at this point.) And if Sapiens were hunting in wider and wider circles with increasingly intricate weaponry — bows and arrows and spear throwers have been suggested by the Harvard anthropologist Ofer Bar-Yosef, but none have been found this early — that competition for game would also have brought bands and tribes into regular and perhaps not so amicable contact, so it is hardly surprising that some means of ornamental identity should have been created.

What the anthropologists do not elaborate is that the number and variety of the ornaments also probably indicate a significant statement of *individual* as well as group identity, just as they do today, and very likely a display of personal attractiveness, a form of peacock's-tailism to indicate beauty and fitness in enticing the other sex. This is significant because, like the beads at Blombos Cave, they indicate a way of perceiving and proclaiming the Self, sending signals of self-worth in a form unknown among earlier humans; though Erectus males must have competed for mates in some way, nothing like this sort of adornment and, well, vanity is found in the Erectus fossil record, where the indications are that group identity was paramount.

It is groups of this developed and increasingly sophisticated Sapiens culture of the Levant — emphasizing hunting (including fishing and fowling), possessing art and ornamentation, having magic and perhaps a sense of the supernatural, knowing individuation — that begin to move northward into the subcontinent of Europe and on to the plains of central Asia sometime around 46,000 years ago. Within the comparatively brief time of 5,000 years, these groups succeeded in occupying a territory stretching east-west for some 4,600 miles, eastward to cover the extent of Europe from modern Bulgaria to Spain, from southern Greece to Belgium, and westward into southern Siberia as far as Lake Baikal, a remarkably swift imposition of one species that may have no parallel on that scale in biotic history.

It seems not to have been climate change that sparked these migrations, since Europe in the period after about 48,000 years ago continued to get gradually cooler and dryer, with a few periods of warmer climate every 5,000 years or so, and would not have been especially more desirable than the Mediterranean-tempered Levant. One break in the coldness did come about 45,000 years ago, when the polar ice cap might have moved as far north as the Swedish peninsula, the ground-frozen tundra grasslands receded to roughly the 50-degree meridian across upper Europe, and a large stretch of open pine-lands in the temperate zone and deciduous forests around the Mediterranean would have been attractive areas of settlement, especially for a species that knew how to adapt to the cold with fire and clothing. Still, it seems more likely that it was the repeated phenomenon of overusing the environment, overhunting in particular as Kuhn and Stiner have suggested, that forced the Sapiens exodus at this point.

There are no sites that record the corridors of these migrations with any precision, largely because paleological work in Turkey has been so sparse. But enough research has been done in Georgia, on the eastern side of the Black Sea to Turkey's north, to show that even in colder periods the area would have supplied reliable and relatively accessible sources of vegetation and game, plus many natural rock shelters and caves, and an abundant supply of flint. This suggests to people like Bar-Yosef that Georgia was on a pathway out of the Levant, possibly one that followed a river like the Euphrates into and over the Taurus mountains and then around the Black Sea into Europe and around the Caspian Sea into southern Russia and the West Siberian Plain.

But since the earliest sites in Europe with toolkits like those of the Levant appear in modern-day Bulgaria, to Turkey's north on the *western* side of the Black Sea, it seems to me more probable that the route out of the Levant would have gone that way. And, to avoid the problem of crossing the mountain ranges of Turkey, it might have followed along the Mediterranean shore — in this colder period, with water locked in the glaciers, the sea would have been smaller and the shore consequently broader — just as we presume was true for the Sapiens who crossed out of Africa along the shoreline of the Arabian Sea. No sites have been found to support this conjecture, because they would be well under water today, but a recent study of the DNA of European populations today suggests that "the earliest migration into Europe . . . took place from the Near East . . . 45,000 years ago" by two routes, one along the Turkish coast and into Greece and southern France, the other across the Balkan Mountains and along the Danube into Germany. And another genetic test of matching Y chromosome markers has found that a small percent of Europeans can "trace their ancestry directly to the Levant of 45,000 years ago."

The chief difficulty in tracing when and where Sapiens settled in Europe — indeed, proving that they *did* settle there — is the paucity of human fossils for the whole period from 45,000 to about 36,000 years ago. German researchers in 2002 found three bones of a skeleton from the site of the original Neandertal discoveries in Germany that they believe may be from a Sapiens female of about 44,000 years ago, and at one of the earliest sites where toolkits of the Sapiens sort have been found — Bacho Kiro Cave in central Bulgaria, dated to at least 46,000 years ago — there are fossil remains of a hominid (fragments of an upper and lower jaw of a juvenile, and a tooth), but it is not indisputably Sapiens and could possibly be Neandertal. The earliest uncontested Sapiens bones in Europe are from Pestera cu Oase, in Romania, at 36–34,000 years ago, and Kostenki, Russia, at 36,000 years ago, and after those any number of sites with bones showing that Sapiens moved into Europe in considerable numbers: Bacho Kiro in Bulgaria, Abri Tapolca in Hungary, Silicka Brzova in Slovakia, Kelsterbach in Germany, Cro Magnon in France, and Kent's Cavern in England.

But in the absence of certain human remains for the long period extending from 45,000 to 36,000 years ago, archeologists can rely on the many sites with tools and artifacts that can be traced to Levantine roots and indicate Sapiens movements into Europe. For example, the toolkit found at an extensive cave called Temnata in the Balkan Mountains of what is now northwestern Bulgaria, dated to about 46,000 years ago, is very similar to that of the

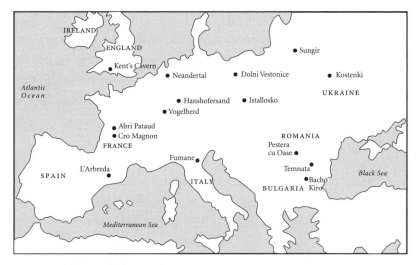

2. Europe, 35,000 years ago

49,000-year-old Boker Tachtit site in the Negev, and archeologists can tell from the way the tools were shaped, even the kinds of knapping blows used on the stones as well as the kinds of tools made, that people of the same modern culture were responsible. The connection is confirmed by similar toolkits found at a number of eastern European sites from Austria to Russia with dates between 46,000 and 40,000 years ago (Istállóskö, Bacho Kiro, Bohunice, Mladeč, Szeleta, Stránská Skála, Willendorf II) that are quite different in pieces formed, and by extension in the thought processes involved, from those of the Neandertal populations who lived at the same time in roughly these same areas. There can be no doubt that this rapid and extensive transfer of technology was a result of successive immigrations of Sapiens, from the Levant and then largely along the major rivers of eastern Europe (Danube, Dniester, Dnieper, Desna, Don), the leading curl of a population wave that was to extend over almost all of the subcontinent of Europe, to northern Italy (Fumane) by 44,000 years ago, northern Spain (L'Arbreda) by 42,000 years, and western France (Abri Pataud) by 40,000 years.

Yes, Neandertals. This species had evolved in Europe from around 250,000 years ago and soon occupied the subcontinent, though probably not in large numbers, from Gibraltar to the Black Sea, from the English Channel to the

Italian boot, eventually extending its range deep into Asia as well, as far as the Aral Sea. Such longevity and expansion indicate that it was clearly a successful species, and yet it did not survive the Sapiens invasion for long; its range was diminished greatly by 35,000 years ago when a second wave of Sapiens moved eastward from central Asia, and perhaps not later than 32,000 years ago, at its last remote refuges in southern Spain and Croatia, it became extinct.

The reason for the demise of the Neandertals is impossible to read from the fossil record. It has been suggested that the Sapiens newcomers might simply have done away with the Neandertals to eliminate the threat of competition for game. Sapiens "were intelligent people, quite capable of sitting around a fire and logically planning the conquest of a region," the English archeologist Paul Pettitt has written, and in places where the two species competed for food, "the attitude may have been to kill first, ask questions later," thus achieving "the modern human race's first and most successful campaign of genocide." Besides, as the archeologist John Shea of SUNY, Stony Brook, has said, "Modern humans [were] very competitive and really good at using projectile points to kill from a distance." But while it is true that Sapiens would go on to hunt other animals to extinction, and in historic times have certainly proved capable of genocide, we can tell by comparing the toolkits that their hunting weapons and strategies would have been so far superior to those of the Neandertals—even Neandertals who may have adopted some Sapiens weapons—as to give them little reason to fear any competition in the hunt. (Because adult Neandertal skeletons consistently show evidence of multiple fractures, particularly of arms and skulls, they must have been doing their killing close-up, which could not have been very often a successful tactic.) Taken with the unlikelihood of there having been much reason to kill these hominids for food, especially as they were so thin on the ground, there would seem to be no particular purpose for the Sapiens to effect genocide.

It has also been suggested that somehow Neandertal women were so strongly attracted to Sapiens men—with smaller, flatter faces they may have evoked the cuteness of children—that they took to intercourse with the newcomers, neglecting their mates, and caused a dire downturn in the Neandertal reproductive rate. Such unions would have been possible, as Sapiens men are generally known to have few inhibitions about coition with strangers, but they seem not to have been productive. Some skeletal remains that certain archeologists read as having features of both species might add weight to this interbreeding thesis, but the general leaning of the paleoanthropological world these days is toward the idea that these "mixed" remains fall within a normal range of Sapiens bodily diversity. As Richard Klein of Stanford puts it, speaking of central European samples, "In overall morphological pattern . . . they are unmistakably modern, with no true Neanderthal features."

Far more likely is that the Sapiens simply out-competed the Neandertals in every valley and steppe where the two species coexisted. They were more numerous and had wider ranges of travel and trade, they were abler hunters, they had a much greater array of weaponry, they had long experience in killing all kinds of large mammals, they presumably had language to aid in coordinating hunting groups, and they had a range of game foods that included fish and fowl (recent examinations of Neandertals' bone chemistry have indicated that their diet was almost exclusively red meat and lacked the beneficial fatty acids of fish). An interesting study of demographic models by the archeologist Ezra Zubrow has found that if there is only a slight decrease in life expectancy in a Neandertal population, on the order of 1 or 2 per cent, and a comparable increase in a competing Sapiens population, it would take only thirty generations, less than a millennium, for the first population to die out. This is the story at every cave site in Europe where the two populations coexisted: the lowest layers are those of the Neandertal culture, overlain by those of the Sapiens with no very great gap in time, and it is never the other way round. "We had become very good at invading ecosystems and exploiting them to the hilt," says Niles Eldredge of the American Museum of Natural History, "sometimes at the expense of competitors."

Just how developed, how powerful, was this Sapiens culture that took shape in Europe in the years from 46,000 to 30,000 before the present and so completely displaced the Neandertals?

At the beginning, as we have seen, the culture looked in many respects like that of the Levant, though with many regional variations in different European settings. But it quickly developed more extensive, more specialized, and more efficient techniques of tool production, greatly enlarged the types and numbers of blades and points, and began to use new materials and develop new rituals. By 40,000 years ago or so it had developed into a distinctive European culture—it is called the Aurignacian, after the cave site Aurignac in southern France—that was clearly descended from the modern culture as it came out of Africa and the Levant but was advanced in complexity and sophistication, in sheer powers of production and procurement, in artistry and ornamentation, as if it had gone through some process of intensification as it explored and settled the new lands.

The climate was dry and cold most of the time—this is the Ice Age after all, with what is calculated to have been an average yearly temperature of 40–45 degrees Fahrenheit—except for those brief thousand-year periods of warmer and wetter weather. This was an ideal climate for extensive grasslands and open coniferous forests, which displaced dense woodlands over vast areas of Europe and central Asia in the middle latitudes, and in turn ideal for a very large variety and teeming number of animals: wolf, rabbit, hare, marmot, and other small game, but especially hoofed mammals that thrive on grasses, including elephant, mammoth, rhinoceros, hippopotamus, saiga antelope, giant deer, wapiti, aurochs, elk, ibex, and, in special abundance, the herd species reindeer, bison, horse, and red deer. And hence an ideal climate for human hunters.

Sapiens had been adept enough hunters in the Levant, and were able to broaden their range of prey animals in times of need, but they never had the abundance of species and the immense numbers of animals—mostly mild, unferocious, and quite large mammals—that they had before them in the wide grasslands of Europe and central Asia. Given the prey, they became predators on a scale and with a skill unprecedented, and blossomed into a culture of complexity and conquest, and with an impact upon their surroundings, such as the world had never seen.

What perhaps most marked these hunting societies was that they made the task of finding game into something of a science. Unlike the Neandertals around them, who are thought to have been "opportunistic" hunters and foragers at this point—killing such game as came by their campsites, scavenging such carcasses as they happened upon—the Sapiens ranged over wide areas and planned ahead to find their prey. They were able to go to the mountains where they knew that the horse or bison herd would be migrating in summer, to the valleys where they would be in winter, and most likely move through a series of camps depending on the season, occupying each for only a few months, concentrating primarily on one species after another. This "food-management strategy," as the University of Illinois anthropologist Olga Soffer has described it, based on "a purposeful 'mapping-on' to resources over a larger region," represents "a qualitative change in both the perception and in the utilization of nature" from those of the Neandertals. And was obviously an important reason for the Sapiens' success.

To exploit this abundant range of game, Aurignacian societies developed what the Cambridge archeologist Paul Mellars calls "a sharp increase in the scale and intensity of hunting strategies," as one can judge from the "sheer abundance" of large-herbivore bones at Sapiens sites. One sign of this is an even more extensive kit of weaponry than their Levantine ancestors had, a great many pieces hafted, with a wide range of functions and forms—different weapons for different animals, it is thought—in particular, Mellars says, "those involving complex, multi-component hunting missiles." For example, at a rock shelter called Riparo Mochi in the Italian Liguria, hundreds of almost identical bladelets have been found, too small to be handheld but perfect as teeth if mounted on a handle of some sort to function as a saw or sword, a new device in Sapiens history. Also, a great many new tools were devised not for hunting but for dealing with the prey after it was killed: long-bladed knives, for example, some with bone handles, to dismember the carcasses, scrapers to remove hair and fat from the hides, "burnishers" to process and smooth them for clothing, and awls or burins to make holes for sewing them together.

Another sign of hunting intensity, again an ingenious response to climatic hardship in the colder regions, was systematic fur trapping. It is not known for certain what the traps would have looked like, but the remains of wolves and foxes in sites in eastern Europe and Russia (Avdeevo, Eliseevichi I, Kostenki, Mezhirich, and Mezin for example) are, in Richard Klein's words, so "extraordinarily abundant" that it is clear they would have been trapped rather than hunted; besides, the close-range killing of dangerous predatory pack animals would have been daunting even for these superior hunters. Most skeletons are whole or nearly so, usually without feet, indicating to Klein that the Sapiens had skinned them with the paws attached, "as modern trappers often do," and turned the pelts into fur clothing.

Stone and flint were used extensively for making points and blades, and here again, as in southern Africa earlier, this kind of material was so important to these hunters that they would travel extraordinary distances to get it, something that the Neandertals never did. At the earliest Aurignacian site, Bacho Kiro, more than half the flint used for artifacts came from outcrops more than seventy miles away from the cave, arguing that the people there were willing to put in a considerable effort—it would mean a journey of maybe four or five days out and back, carrying heavy rock one way—to have high-quality weaponry. (One reason to put the effort into high-quality flint is that its properties can be enhanced by heating and slow cooling, or "annealing," to produce a more elastic surface that can be knapped and shaped more easily.) At the Czech site of Dolní Vestonice, just a little south of modern-day Brno, people apparently found the local chert not good enough for knapping, though elsewhere it was often used, and they were willing to go to quarries 120 miles to the north, in southern Poland, to get high-quality flint to use for over 90 per cent of their tools; they also collected radiolarite of red, yellow, and olive from 100 miles east, and even some obsidian from outcrops 100 miles south in Hungary. It is not likely that such a wide area could have made up the Dolní people's exclusive hunting territory, so it is probable that they encountered other similar people on such long excursions, perhaps even trading with any who occupied the favored quarry sites. The necessity, and ability, to make bonds and alliances with other people in this way would also have made it possible to call upon them in time of need—during a local drought, for example, or a severe winter—and to return the favor when asked; this sort of reciprocity is found in many historical tribal settings, as with the !Kung of the Kalahari and the Loikop of Kenya.

Despite the obvious importance of stone, what is especially distinctive in this Aurignacian culture is for the first time the widespread use of bone, antler, and ivory—from mammoth, oxen, rhinoceros, reindeer, red deer, bison, bird, fox, hare, from almost any animal that could be caught or scavenged—and for dozens of tasks ranging from scraping to cutting to piercing.

But above all for hunting. Spear points were now routinely made out of these animal by-products, and in great numbers and at almost every site, a refinement that is rare in the record after the initial appearance of bone points in southern Africa. Working on these hard surfaces was time-consuming and difficult, far more so than knocking off blades from a piece of flint, and the allotment of time to this rather than other more immediately beneficial pursuits like foraging or scavenging—or reproduction itself, for that matter—indicates the extraordinary emphasis placed on efficient hunting. Antler and bone would have been of special importance, it has been suggested, because they would be light enough that several spear points could be taken by each hunter in a band, and they would be easier to resharpen if the tip broke off when, say, hitting an animal's bone. They were also easier to shape, for a firmer and more durable attachment to a shaft—antler and even ivory can be made quite soft and pliable by soaking in water—and a reliable spear would have been an essential weapon for hunting big and powerful animals like bison or mammoth without getting too close. A distinctive kind of spear point, usually of antler but sometimes of bone and ivory, with an upside-down-V-shaped split at the base for a more secure fit to a pole or handle, was in fact developed early (the earliest at Istállóskö, Hungary, about 44,000 years ago) and became a hallmark of the culture for the next 17,000 years. Reindeer was the chosen species, for both sexes have antlers (the average male has a rack fifty inches long with twenty-eight points) and it is found throughout Europe as far south as the Pyrenees: fifty antler spear points have been found at one level of Abri Blanchard alone, seventy-five at Isturitz, both in southern France.

But also for ornaments. Just as in the Levant, ornamentation was an important ingredient in the lives of Sapiens in Europe, and here it may have played an even more decisive role, for now they used antler, bone, and mammoth ivory for their beads and pendants, plus animal teeth (fox mostly), in addition to marine shells. Evidence for this comes right at the start—pierced fox and wolf teeth from Bacho Kiro at 43,000 years ago, pierced animal teeth from El Pendo in Spain at 40,000—and continues down through the Aurignacian period and beyond throughout the subcontinent. And at many sites with such elegance and in such quantities that it must clearly have played some essential role: "For all modern humans," writes the anthropologist Randall White of New York University, a specialist in Stone Age ornamentation, "personal adornment is one of the most powerful and pervasive forms by which they represent beliefs, values, and social identity."

One simple indication of the key role of ornaments is that from Spain to Russia beads and pendants were made out of almost everything: steatite, limestone, schist, hematite, pyrite, lignite, talc, belemnite, jet, soapstone, and coral, plus teeth and bones and ivory from mammoth, red deer, beaver, moose, bison, fox, and hyena, *plus* shells of a score of marine animals (periwinkle favored in France) from the Atlantic, the Mediterranean, the Black Sea, and all the major rivers, and even fossilized shells that had worked their way to the surface. Another is the sheer profusion of these ornaments—"the explosion of items of body adornment" is how Randall White describes it, citing in particular a trio of cave sites in southern France (Blanchard, Castenet, de la Souquette, dated to 34–32,000 years ago) where he found 835 beads of ivory and stone, plus countless shells—even though each one involved "a complex production sequence that includes piercing, grinding, and polishing," the last with the use of powdered ocher. And finally there are the extensive distances over which material for beads was transported, as at Kostenki, on the Don River, where mollusk shells came from the Black Sea coast, 300 miles south, and Pavlov, in the Moravian region of the Czech Republic, where the shells are thought to have come from the upper part of the Adriatic Sea, at least 350 miles distant and a with a stretch of the Alps in between.

The importance of ornamentation obviously went beyond just individual display or group identity: it now played some kind of a social or political function, establishing roles and assignments for band and tribal members. White argues that decorations of this period not only convey "complex systems of meaning and social action" but also show evidence of "new kinds of social systems," presumably more intricate and developed organizations that were necessary to secure internal cohesion and cope with the stress and tension of hunting in ever-colder climatic conditions. Whatever form these new systems took, their very necessity indicates that some kind of severe pressures were at work: could it be that the strains of maintaining a regular domination over so many other species—more time spent killing and using more kinds of animals, a life increasingly steeped in blood—were beginning to exact a deep mental and emotional toll?

In any case, one more new and potent cultural phenomenon also pointing to the effects of environmental pressure appears in Sapiens society a little after the widespread use of ornamentation. It is the extraordinary outpouring of art—sculpted, engraved, drawn, and painted art—beginning around 35,000 years ago (Chauvet Cave, the oldest, is dated to 36–35,000 years ago) and continuing throughout Europe over the course of some 25 millennia, work that still strikes us all these eons later with its beauty, power, emotion, and sophistication. Of course there were earlier indications of the Sapiens capacity for art, from the ocher pieces at Blombos Cave to the slashed horse at Hayonim, but now there is such an increase in abundance and in proficiency that it rises to a new level of achievement and marks a new degree of its importance to these societies. ("Art," to be sure, as we would regard it today, though its creators would have had no such concept.)

Different kinds of art were put to different purposes, some of which we can understand today, but all seem to have been used for one form of magic or another, images that would be manipulated to advance some human end. The little sculpted figures, for example, found in great profusion in this era throughout the Sapiens range from western Europe into central Asia (and especially along the Danube and its tributaries) depict a full gamut of animal and human forms, prey animals of all kinds (though mammoth preferred), some fierce animals, women and female body parts, and men (rarely) and male phalluses (often). Most of the animal figurines are thought to be examples of totems that could be carried around and used in hunting magic, since many show signs of having been used repeatedly—such as the ivory carvings of bison, horse, mammoth, and reindeer at Vogelherd in Germany that are "marked and overmarked as though in periodic ritual," according to the researcher Alexander Marshack, presumably by imitation spear and knife marks to prosper the hunt; the sheer number and ubiquity of the pieces— one estimate is of "tens of thousands"—accords well with the importance of

6. "Lionman," ivory, Hohlenstein-Stadel Cave, 32,000 years ago. Copyright Ulmer Museum. Photo by Thomas Stephan.

the hunt in Sapiens' lives. Some other sculptures, especially in the Moravian area of central Europe and the upper Danube in southwest Germany, depict carnivores—in particular bears and lions—and though these animals were not usually hunted, they would have been the subjects of different kinds of magic, say to dispel their danger, or harness their power.

One of the most unusual figures, of yellowish mammoth ivory, nearly 11 inches high and almost 2½ inches round, comes from the Hohlenstein-Stadel Cave in southern Germany and is dated to around 32,000 years ago, making it one of the oldest. It has a standing human body with prominent shoulders and heavy upper arms, and the head of something catlike, usually regarded as a lion, with a slight, almost all-knowing smile. It is commonly called a "sorcerer" or "shaman," because magic-makers in almost all historical tribal societies wear masks of animals in their rituals, and it might have been used to create a shamanistic spell, common in the ethnographic record for controlling the behavior of animals, healing the sick, and changing the weather. Not all scholars accept the existence of shamanism this early in the human record, and it is possible that the piece was merely meant to endow the human holding it with the strength and ferocity—and smugness—of a lion, but in any case it had a clear magical purpose.

Not all the Aurignacian sculptures involved hunting magic—as many as two hundred of them, bulbous, naked women with exaggerated breasts and stomachs often called "Venus figurines," clearly were used for some kind of reproductive or fertility magic—but that appears to have been the purpose for the great majority of them. Nowhere more spectacularly than at two adjacent hillside sites in Moravia (Dolní Vestonice and Pavlov) occupied between 28,000 and 27,000 years ago, where there is evidence of the first instance of making ceramics, not for pottery but for a weird and perhaps unique form of hunting magic. Two open-air kilns have been discovered at Dolní Vestonice, dug into the earth, with thousands of clay figurine fragments inside them, more than 6,700 in all, and nearby areas have yielded another 4,000 or so, the remains of what had originally been perhaps as many as 3,700 clay figurines (98 per cent of which were animals), molded from the local hillside loess, then fired—and *exploded*. The fragments under a microscope have rough, branching edges, not at all like pieces that are smashed apart and weather to a smooth edge but rather like those that are produced in a kiln by a "thermal shock" explosion when clay not completely dried is fired to a high temperature. They *all* have these rough edges, all 10,000 plus: "Either we are dealing with the most incompetent potters the world has ever seen," says Olga Soffer, an expert in this region, "or else these things were shattered on purpose."

That purpose was not likely to have been some kind of fireworks entertainment, and the destruction of the figurines would seem to have involved too much work (perhaps as much as forty hours per figurine) for just an expression of general anger at the local fauna; it was rather an example of some sort of magic that would have special meaning for the hunters collected there— destroying the token "soul" of an individual animal figure so that its real-world counterparts would be vulnerable in the hunt, or giving hunters a sense of inevitable power over the counterparts in the future. It seems to have been "carried out by only a small number of people," Soffer thinks, who had special "control over this behavior"—shamans, perhaps, or special hunting masters—and the obvious amount of time and effort taken to create and demolish the figurines, a completely nonutilitarian (not to mention nonproductive) activity, suggests that it must have had some high significance.

And in the famous European cave art, in at least 350 caves with what one researcher estimates to be 15,000 paintings in all, we find evidence of the same sort of underlying magical purpose and evidence of its critical importance in sustaining the tribal societies. At least 15 per cent of the animals overall have markings and lines that seem to represent spears, as with the Hayonim Cave tablet, which can be taken as simple hunting magic, and of course other animals (the great majority of which are prey animals) could have been used for the same purpose in rituals that did not leave marks. A great many half-human, half-animal figures are depicted, again either shamans or humans taking on animal qualities; most caves have figures with magnified genitalia or U-shaped lines taken to be vulvae, and some even show pregnant or copulating animals (as at the famous Altamira Cave with its mating bison), presumably to assure the continued fertility, and availability, of the favored animal population.

No one who has visited these painted caves has ever doubted that there was ritual significance to the depictions, and it has even been suggested that they were used for manhood initiations and vision quests. Certainly the effect of being in a deep cave—underground and away from the familiar; damp and silent and cold; stalagmite concretions rising eerily from the floor; all a palpable darkness except for the flickering torches and animal-fat fires in small stone dishes, where whispers echo off the rocks and around the huge chambers—could not help inducing a sense of the awesomeness and mystery, and the fearsomeness too, of nature. Paint a valuable animal there, with realistic foreshortening and perspective and shading so that it seems almost alive, especially as the waving shadows move across it, and it is likely that anyone who comes to it would feel a shiver of recognition in the presence of forceful living beings, where such a thing as hunting magic would seem as possible and effective as slicing meat. A contemporary scholar has described the feeling: "The animals become animated in a flickering, yellow light which plays on the hollows and projection of the cave walls. Sometimes they seem to be moving together deeper into the cave or toward the entrance. Sometimes the flame gives a curious effect of imminence, a different kind of movement. The animals are not going anywhere. But they seem alive nevertheless—breathing, relaxed or tense, and ready to move." If someone today can feel this, imagine how much more the humans of 30,000 years ago would have felt it, those who lived in small bands in a wilderness surrounded by innumerable animals, who depended on the successful hunting of those beings for survival. In that intensified experience, human fear and insignificance could easily be transformed into human might and meaning.

And that is what Sapiens art is all about. Whatever kinds of magic and ritual were practiced with the sculptures and paintings, of which we can have only a small idea today, they all involve some form of human effort to have control over nature, *to extend human domination*, as Freud and Frazer long ago pointed out but the anthropologists too often forget. With symbolic art, and particularly the charged rituals in deep caves, humans became involved in a new relationship to the animal world, or at least were attempting to extend their old relationship in a new way. The paleontologist Niles Eldredge sees this extension of "power over the natural world in general" as "a first step — and a necessary one — to declaration of full-scale independence from the natural world." How fateful, that: the attempt to be *independent*, or to think of oneself as independent, from an ecosystem on whose bounty one is entirely dependent for sustaining life itself is delusional, and can be maintained only by tortuous ideas of self-importance and wrathful practices of self-enhancement. The Canadian naturalist John Livingston suggests just how consequential the invention of magic was: "Man had ceased to be an integrated part of the natural ecosystem. He now had something which was even more important than flints and spears. In its significance, it ranks only with fire. René Dubos has remarked that the new magic was 'probably more important for the understanding of man than physiological and biological knowledge of bodily structure — indeed, more important than the development of tools.' From here on it was the modern era of human dominance."

Or, as I would rather put it, the modern era of human dominance, which had really begun some 35,000 years earlier, was entering a new and momentous phase.

The essential question to ask of this phenomenon is, of course, why? And why now?

Almost all scholars in this area are agreed that this widespread and fairly rapid emergence of art had to have been in response to some extraordinary kind of new pressure, yet there seems to be no agreement on what this might have been. But as I suggested earlier, I think it is possible to identify two serious and sweeping phenomena around 35,000 years ago that would have created conditions under which people might be forced to create rituals around little ivory sculptures and animal cave paintings.

For one thing, the climate suddenly started to grow much colder after 37,000 years ago, at the end of the warmer "Hengelo interstadial," as Europe began to enter what is termed the "full glacial" period. True enough, Sapiens had known long periods of cold during the Ice Age, but the oxygen-isotope record from deep-sea sediment cores indicates that this onset was particularly abrupt and severe. It was accompanied by a decrease in precipitation of maybe 15 to 20 per cent (in some places 50 per cent), and dry summers and winters that would have severely limited plant growth, for both humans and animals. And as the polar ice sheet moved deeper and deeper down into northern Europe after about 35,000 years ago, some populations where game species such as reindeer persisted may have made local adjustments (fur clothing, animal-hide shelters), but most would have had to move south.

Only two regions provided conditions sufficiently benign to attract large numbers of big game and the people dependent on them: the European Southwest, especially in southern France and along the Bay of Biscay in Cantabrian Spain, and the Russian plain above the Black Sea in the Southeast, especially in the Dnieper and Don river valleys. Here, though the temperatures were as much as 10–12 degrees colder than today, there were abundant tundra grasses and even trees like oak and elm in the sheltered valleys. Large numbers of arctic-adapted animals—mammoth, reindeer, and horse most prominently, but also some woolly rhinoceros, bison, red deer, and giant deer—moved into the areas, feeding on the grasses, though there were certainly temporal and regional shortages of one species or another throughout the whole period. People followed the animals—those at any rate who acted fast enough to escape the long and famishing winters to the north—and settled into river valleys protected from the sweeping glacial winds where big game came to drink or pass through on their annual migrations. Most hospitable of all were the deep, narrow valleys along the rivers of southwestern France—the Garonne, Dordogne, and Lot in particular—where settlers could take advantage of the numerous, deep, south- and west-facing caves and rock shelters and could prey upon the animals that came down the narrow riverine corridors.

An increasing number of people in these favored regions meant an increase in competition for the same wild animals, so the hunt would take on even more importance than before and the need for a regular supply of game would be paramount. In such conditions a new and intensified form of hunting magic such as art provided would be nearly inevitable.

And the competition was further exacerbated by shortages in game. Some species almost certainly declined in numbers and in average body size, for their new areas generally provided smaller ranges and shorter growing seasons for their foliage, and species that could not adapt to the increasingly cold-dry climate became regionally extinct. Certain types of warm-adapted rabbit and deer gradually died out after about 30,000 years ago, as well as one variety of elephant, the woodland rhinoceros, and the European hippopotamus, and even the number of species of cold-adapted mammals went down by something like a half (from fifteen to eight or so). This would have added to the competitive pressure on the hunting bands and increased the need to develop new rituals for expressing and transmitting power over the favored prey, for which painting in particular would have been a natural medium and deep, emotive caves a natural gallery. That might explain why so many of the painted caves of Europe, *roughly 85 per cent*, are in the Southwest, where the population densities and stresses would have been greatest.

A second process, perhaps even more important, accompanied and magnified the effects of climate stress at about this same time. According to the population geneticist Spencer Wells in his new analysis of the geographic distribution of Y chromosome markers, most of the men in the European population of today can "trace their ancestry back to central Asia within the past 35,000 years." As he reconstructs it from genetic studies, the people who had migrated out of the Levant into the steppelands of central Asia after 45,000 years ago first dispersed eastward into Siberia around 40,000 years ago and then westward, perhaps forced by the deteriorating climate off the open steppe and its fierce winds and frozen turf to go in search of sheltered valleys and warmer environments. They traveled across the mammoth steppe of Russia and the Ukraine as the game herds moved west, then along the major rivers of central Europe, and into the Atlantic provinces of western Europe around 35,000 years ago.

A human migration of this magnitude—the preponderance of the Asian gene in Europe suggests large numbers in the original sweep—would have greatly modified European society in a few thousand years. Presumably the newcomers absorbed the earlier Sapiens populations that had come along the Mediterranean route, displacing the Neandertal populations in their path, and merged their culture with the Aurignacian, from which it would not have differed much in any case since they both arose from the same Levantine roots; this process would account for the variations in Aurignacian toolkits at different geographical locations and for the increasing number of sites with evidence of Sapiens occupation in this period.

Add this population influx at a time of growing food scarcities, and it is obvious that human societies would have had to take great measures to survive, and again art can be seen as a reasonable ritualized response. That this was a particularized response to the great stress of maintaining the hunt under severe climate and migration pressure is further indicated by the presence of very little portable sculpted art, and a total absence of cave paintings, at this time in other parts of the world—the Levant, for example, or northern and southern Africa—where such pressures did not exist. Ofer Bar-Yosef has pointed out that "imagery is related to ritual," and the rituals developed in western Europe were unique, a product of a particular kind of stress that led to "the particular social structure of groups, i.e., the intensity and frequency of social interaction on all levels." In the Levant, by contrast, where there was abundant plant and animal food throughout this colder period—"this region was more lush than many other parts of the Old World and well suited for continuous habitation of human groups"—there was no such stress and hence no recourse to art; it was not until later, after 13,000 years ago, that the area experienced population pressure and the art of the Natufian culture was created in reaction to it.

Art in whatever form, and the rituals that attend it, imply something like religious belief, if we can use "religious" in the broadest sense to include a deliberate use of symbolic ceremony invoking other forces to satisfy human wants and needs. And so it is just now, a little after the introduction of cave art, that we find unmistakable evidence of deliberate human burials for the first time, accompanied with ornaments and artifacts, sure evidence of some belief in an afterlife, and hence a belief in what could be considered a human "soul," and hence what we can only call religion.[5]

But reflect on the rite of deliberate burial. As noted several times above, it seems not to have been a practice of earlier humans, certainly not a common or widespread one, and indeed there is no sign of it during the long eons of *Homo erectus*. It appears now, with the climate and population stress, when it is probable that death became an event of oppressive regularity for the people of Europe and the desire to negate it in some way must have grown as a natural reaction. For people to deliberately inter a body in a special place underground and decorate it with treasured goods they will never see again, they must have a powerful idea that they can in some sense supersede death, or at least provide another life beyond this one. It is a refusal to accept the inevitability of death, a desperate denial of the fate that faces the members of every species, and it argues a social psychology at odds with the inviolate ways of nature, which is not a healthy state: to refuse death is to defy the natural.

The earliest graves, with bodies carefully placed and grave goods alongside, often covered in red ocher dust, occur in central Europe around 30,000 years ago, and from that time onward burial sites are found throughout the subcontinent. Not all tribes practiced formal burial, as near as we can tell, and not all members of a given tribe would be accorded that honor: it was clearly a ritual often reserved for people of special standing, and the number and kinds of grave goods would indicate the status. Some graves are fairly simple—the various levels of the Grimaldi Caves after 25,000 years ago in Italy show figures with simple headbands and armbands of worked bone— but others are quite complex, like the male skeleton, dated to 23,680, in the Moravian city of Brno, surrounded by a mammoth shoulder blade and tusks, ribs and a skull of a rhinoceros, horse teeth, more than six hundred fossil shells, two large, perforated stone discs, fourteen smaller discs of stone, bone, and ivory, a polished reindeer antler, and a male figurine of ivory, and all of it covered in red ocher dust.

But the most striking burial of all is at a large, open-air site called Sungir on the banks of the River Kliazama, about 125 miles northeast of modern Moscow and as far north as humans had ever settled to that time. There, about 28,000 years ago, eleven people were buried in the frozen permafrost, probably not all at the same time but within years of each other. Eight of them, represented by only fragmentary remains, look to have been deliberately interred, although there is no sign of grave goods or ornaments, but three of them—a man of about sixty, according to the Russian excavators, and two children, one possibly male and about twelve years old, the other possibly female and eight or nine—were given the most spectacular send-off known to prehistory.

The man, stretched out on his back with his hands together at his pelvis, had been dressed in fur or leather garments—antler needles are known from at least 26,000 years ago—that had been painstakingly decorated with 2,936 small, round, pearl-sized ivory beads in dozens of strands; around his head in the shape of a crown (possibly a cap) were additional strands and a number of arctic fox teeth. Around his neck was a flat pendant made of schist, painted red with a small black dot on one side, and on his forearms and biceps were twenty-five large mammoth-ivory bracelets, each a quarter of an inch or so wide, polished and decorated with red and black paint. What is remarkable about these bracelets in the estimation of Randall White of New York University, who has made a careful analysis of the site's artifacts, is that they must have been sliced lengthwise from a tusk and then presumably boiled for some time to make them malleable, then forced into a circular ring and fastened together through two small perforations at each end.

The putative boy, buried full length and similarly supine, head-to-head with the girl, was decorated even more elaborately, with 4,903 ivory beads, slightly smaller than the man's, a similar crown-like arrangement, rings of beads below his knees suggesting decorated boots, an ivory pendant in the shape of an indeterminable animal, an ivory fastener at his throat, and what was presumably a belt decorated with more than 250 perforated fox teeth at his waist. The girl was grander still, with 5,374 beads the same size as the boy's, a similar beaded crown, and an ivory fastener at her throat.

But that's not all. Near each of the bodies were grave goods that included weapons presumably intended to accompany these three honored people

7. Adult skeleton, Sungir, 28,000 years ago. Courtesy of Jan Jelinek, from *The Evolution of Man* (New York: Hamlyn, 1975).

into their afterlife and an odd assortment of what could be purely decorative pieces. Beside the boy was a polished human thighbone shaft—it is rare to find human bones used as art, or even as a trophy, in the Stone Age—filled with red ocher; under his left shoulder was a small, flat sculpture in ivory of a mammoth, perhaps a pendant; nearby was a flat, horse-shaped figurine in bone decorated with red ocher, and with two rows of small, drilled perforations following the contours from head to tail and painted black.

Alongside the girl were two antler pieces that have been called wands, or batons, one of them with perforated rows like those on the horse, and several small shafts that have been described as lances, one of which had been inserted into the central hole of a circular ivory disk with a ring of eight other holes around it in a classic rosette pattern. Beside each figure were two hefty spears made from a mammoth tusk—perhaps straightened by boiling, White suggests, or split off and shaped, according to the Russian excavators—the bigger one, about eight feet long, weighing about forty pounds. That would be a heavy load for a single man to wield, and White is inclined to call the spears ceremonial, but it is certainly conceivable that when in use they too could have been thrust into a circular disk like the one with the girl's lance, enabling at least two men to hold them by the outside holes for a powerful thrust. (Then too, the disks may have been merely decorative, like one the size of a quarter with five holes found beside the girl's head, and the one on the lance merely accidental; the rosette shape would be a familiar one in certain later cultures—Egypt, Crete, Mycenea—where it is thought to have represented the sun and was used as a burial ornament to signify rebirth.)

That the burial ceremony was of marked importance in the Sungirian religious system is proven by the beads. There were 13,113 of them in all, each individually crafted in a process that according to White would have taken more than an hour apiece, or at least 546 days, one and a half years total, or four and a half years of eight-hour days. That is a stupendous investment for any society, more so for one in the shadow of a polar glacier whose members had to wrest a living from the frozen tundra and survive the arctic climate. To bury that many carefully crafted beads, along with an array of arduously worked hunting tools, where they will never be seen or used again—at least not by the living—could be countenanced only by a society that had a deep need to honor its favored people and a sure spiritual belief in an afterlife, and presumably resurrection in it.

For the three figures were extraordinarily favored, far above their buried comrades. This is the earliest sure evidence of a system of *hierarchy* in Sapiens society (the earlier use of ornamentation suggests but does not prove it), and it is confirmed in quite a number of other European gravesites around this time (Grotte des Enfants in Italy, for example, La Madeleine in France, Mal'ta in Siberia), though there is no way to know whether it was adopted by all European tribes. What is remarkable about it at Sungir is that it seems to have been based on heredity rather than accomplishment, since the children, who each had more beads than the old man, would likely not have had time to *achieve* special status, especially above his; alternatively they may have been sacrificed to accompany the man into the afterlife, but then why they should have been so copiously decorated unless they had some status is something of a mystery.

The existence of hierarchy and social stratification this early in the record—some 25,000 years before the familiar empires of the historical era, when such divisions have traditionally been thought to have originated—marks a fundamental and far-reaching reordering of Sapiens society. Though it is true that most primates have hierarchical systems, the general assessment of *Homo erectus* society is that it was generally egalitarian and was so for perhaps a million years, much on the lines of a number of contemporary and historical hunter-gatherers, and that this is the way Sapiens also lived for most of their evolution: as Christopher Boehm put it in his recent study of "egalitarian behavior," earlier humans "lived in what might be called societies of equals, with minimal political centralization and no social classes." Sheer survival must have been extraordinarily difficult in this later period, and tensions within the group must have been disharmoniously severe, for people to have changed such a settled and successful system and succumbed to a mode of the favored and disfavored, and to have celebrated and honored it within its core religious beliefs.

John Pfeiffer has argued that stratification was probably inevitable in societies where there was severe population pressure and regular competition, perhaps open conflict, over game animals. The successful hunt became of paramount, indeed life-and-death, importance, and that required two new arrangements: first, "a new division of labor: an elementary two-level hierarchy of people of greater and lesser status, people concerned mainly with planning and control, and people concerned mainly with getting the work done"; and second, the elaborate and mysterious rituals of hunting magic led by a few "masters of the art of deception and illusion," the shamans and sorcerers: "What probably did most to widen the prestige-and-power distance between the few and the many . . . was the explosion of ceremonies, including those held in the art caves." Together with the tensions of dealing with neighboring bands daily, in some cases coming together for ceremonial meetings two and three times a year, "everything seems to have conspired to speed the passing of egalitarian traditions and the rise of status, thus laying the groundwork for future power struggles."

It all does not bode well for human society.

THREE

Intensification and Agriculture

20,000–5,000 YEARS AGO

It is called the Last Glacial Maximum, but that is too arch and scientific a term to describe what must have been the most wrenchingly difficult millennia for the human species since the volcanic winter 50,000 years earlier.

Temperatures everywhere had begun a slide after the sharp dip about 37,000 years ago and reached their coldest climax about 20,000 years ago, beginning a long period of devastating, dry cold lasting for perhaps 2,000 years, broken only by intermediate periods of warmer weather lasting a few centuries or less, followed by another 4,000 years of relative cold. With temperatures as much as 20 degrees colder than today, Sapiens society was challenged as it had not been challenged before, particularly in Europe and central Asia. Huge ice sheets, a mile and a half thick at points, covered the whole upper third of Europe to the middle of modern Germany and Poland, with glaciers as well on the Alps and the Pyrenees. Ice shaded into polar deserts without vegetation (or, eventually, animals or people) in northwestern Europe, northern Russia, and northern Siberia, and below that a huge swath of arid steppe grassland over permanently frozen subsoils, stretching from the middle of France to the Ural Mountains. The two principal areas of comparative hospitality, as they had been since the "full glacial" started 37,000 years ago, were the river valleys of the Russian plain and southern France and Cantabrian Spain, but even there the climate was harsh: winters on the Russian plain, for example, were marked by temperatures hovering around zero and wind chills of –20 to –50 degrees, more like northern Siberia today, and vegetation other than grasses and forbs was meager, though roots and tubers survived in places as well as occasional stands of oaks and firs.

Year after relentless year people were forced from all over Europe into these refuge areas as the cold made vast regions uninhabitable. Some sense of the population pressures there—"packing" is the anthropologists' term—comes from figures on the number of cave sites lived in by Sapiens in southwest Europe in the whole period from earliest occupancy around 40,000 years ago to the end of the Stone Age around 10,000. Whereas only 13 per cent of the total of 169 excavated sites in the Cantabrian region were used in the early years of that period when the Aurignacian culture flourished from 40,000 to 28,000 years ago, 77 per cent were used in the ten years after the glacial maximum; similarly, of the 243 known sites across the Pyrenees in the Perigord region of France cut by the Dordogne and Lot valleys, 18 per cent were occupied in the Aurignacian era, 47 per cent after 20,000 years ago. This suggests an enormous increase—threefold to sixfold—not only in numbers but in density as well, leading to increased contacts among groups, increased tensions, along with increased competition, and possibly conflict, for the available game.

Four cold-adapted herd species now provided the main source of food in the favored regions—reindeer, bison, horse, and red deer, supplemented by mammoth, especially in the East—and though their numbers were large enough in most years, the pressure on them from the numbers of people forced into these regions (neither of which was more than 40,000 square miles) was great. One study, by two Russian scientists at the Zoological Institute of the Academy of Sciences in Moscow, calculated that on the southern Russian plain at least (and southwestern Europe would have been similar) the average person would need 4.4 pounds of meat a day. Figuring a base population in any given area of 5,000 people, that would require 4,015 tons (8 million pounds) of meat a year (about 22,000 pounds a day), which it would take some 36,000 horses a year (at 220 pounds each after butchering) or 12,000 bison (at 660 pounds after butchering) to satisfy. That would mean successful hunts in each region would have to bring down the equivalent of about 100 horses or 33 bison *a day*, every day of the year.

Of course this load would be eased by scavenging, for example at swampy sites where elephants and mammoths are known to have died regularly in some numbers in several places both east and west, and by adaptive hunting strategies like killing smaller game and exploiting fish and fowl. But still, taking as an example the Perigord region, an area of perhaps 10,000 square miles, the 114 sites occupied after the glacial maximum imply the existence of perhaps 100 bands (since some of the sites may have been half-year or temporary settlements). If each band had between twenty-five and fifty people, as many anthropologists think likely for this era, each would need the equivalent of a horse or half-horse every day (50 people would require 220 pounds of meat, or one butchered horse, a day), or a total of between 50 and 100 horses for the region. Every day, all year, year after year. It is not hard to see how that could put a significant strain on the faunal resources of the area, not to mention on the relationships among the bands out hunting for them.

Given the increasing population densities in the severe cold climate and the need for a large number of animals to supply their basic food needs, people had to come up with new strategies, and the anthropologist Olga Soffer has pointed to the general scholarly agreement that "some significant changes can be observed around the last glacial maximum" both east and west as a result of the new conditions. In the west, as the anthropologist Michael Jochim of the University of California, Santa Barbara, has described it, the effects of the "packing" of so much of Europe's population into a relatively small habitable region produced not uniformity and simplicity of lifestyles but "a period of growing complexity" and "social and demographic stress," reflected in a great increase in the number of occupied sites, a much greater "intensification of big game hunting," an expansion of diet to include many more smaller (and difficult-to-hunt) animals, new and finer weaponry for the hunt, a more "elaborate material culture" of cave art and ornamentation, and more complicated interband and intertribal relations to smooth over conflicts arising from competition for game in a constricted area. Much of this was duplicated in the East—aside from the cave art, since there were virtually no caves in that region and almost all recovered artifacts are from open-air sites—where, as Soffer puts it, "climatic stress [due to] deteriorating environments around the glacial maximum would have been more acutely felt."

Of the "significant changes" of this era, the first and most obviously necessary was an increased proficiency in the hunt, marked by new weapons and new strategies. Not that these people were inept hunters to begin with, because they had survived all these millennia with capable technology and accomplished artistry, but the twin necessities of killing enough game and avoiding tribal conflict demanded novel, or at least augmented, approaches. Two paleoanthropologists who have studied hunting systems argue that "a kind of 'subsistence threshold' seems to have been crossed for the first time . . . about 20,000 years ago," when "change becomes much accelerated as a pattern of linked diversification" (more species being hunted) "and intensification" (more effort at hunting), prompting a "redoubling [of] efforts to obtain food despite increased costs" of time and effort.

It was in western Europe that most of the new weaponry seems to have been developed. Many sites show a profusion of new stone weapon points, finely and delicately made: one source says they are "the finest examples of flint workmanship" in the whole period after 45,000 years ago, some with "shoulders" cut into them or stems at the bottom for easier hafting, some "leaf-shaped" of various sizes; the larger points are thought to be for spears and the smaller for darts. There is an extraordinary diversity in these points —anthropologists can make out a dozen different styles, from "Asturian concave-base" to "le Volgu 'monster' laurel leaves"—suggesting that there were a great many different bands living in this area, having a more or less common culture but developing their own distinctive styles. As time goes on a long point known as a "backed bladelet"—a triangular blade about three inches long with one lateral edge blunted—becomes most common (at some

8. Bison with arrow marks, Niaux Cave, 14–13,000 years ago. Courtesy of
Antonio Beltran, Monografias Arqueologicas.

sites in Cantabria up to 70 per cent of the points are of this style), because
it could be fastened to the end of a spear and then easily replaced if it broke
off in an animal's flesh.

These stone points were immeasurably enhanced by a remarkable invention appearing at about this time, the spear thrower. This was a rod of bone,
antler, ivory, or sometimes wood, about a foot long, with a notch at one end
into which the base of a spear (or a dart) could be fitted, so that the hunter
could get extra leverage, and thus extra power and distance, in throwing the
spear; modern replicas using stone and ivory points have been shown to penetrate even a mammoth skin and kill animals the size of a deer at a distance of
twenty yards or more. The oldest known spear throwers are from the Combe
Saunière and Le Placard in the Dordogne region of southern France of about
18,000 years ago, and their efficacy was so apparent that they became widely
adopted in most of Europe by 16,000 years ago; the device must have been
transmitted as far east as southern Siberia before 15,000 years, for it appears
in North America among the Eskimos, where it is called an *atlatl*, and the
earliest migration there is thought to have been around 14,000 years ago.

Another invention most probably in use around this time—for the necessity of survival in these difficult conditions was a very fertile mother—was the
bow and arrow. The bow's advantages for the hunter are clear: it doubles the

range of the spear, it is lightweight and easy to carry, it is easily and quickly re-loaded with projectiles that can be carried in bulk, it can be shot from a variety of positions, and it can be sighted at the target along the line of the arrow for greater accuracy. If it was not in fact invented as long ago as the volcanic winter in southern Africa—where the abundant small points may have been hafted not to arrows but simply to darts thrown or blown—it most probably was in existence in Africa around 20,000 years ago, judging from some bone rods that seem to Richard Klein of Stanford to be arrow "foreshafts." Of course "once the bow and arrow were invented," he says, "they would obviously have diffused very rapidly, and a sharp increase in the abundance of tiny backed bladelets in both Africa and Eurasia" from about 21–20,000 years ago is fairly suggestive evidence of their spread. Barbed points from the site of Parpallo in eastern Spain, for example, dated to 19,900 years ago, have a size and shape that would, as the archery expert Christopher Bergman puts it, "fit comfort-ably into the ranges" of arrowheads known to have been used in historic times, and there are any number of similar points from other sites, including the "shouldered" and stemmed tips, that modern tests have shown could be used in arrows. Even more convincing to me are the depictions of arrows in cave art of this period, as at Cosquer, "the cave beneath the sea," between 19,000 and 18,500 years ago, where unmistakable barbed and feathered designs are engraved on horses, ibex, deer, and seals. (A number of other caves—Coug-nac, for example, on the Lot, at maybe 18–17,000 years ago, and Niaux Cave, in Pyrenean France, at 14–13,000 years—also have paintings with obvious arrow designs on them.) The oldest sure evidence, however—wood does not often fossilize—comes in the form of bow fragments of pine heartwood from about 12–10,000 years ago in France and northern Germany, and some one hundred notched arrowshafts from Stellmoor in northern Germany of about the same age.

It may be that some of the barbed "arrows" at Cosquer in fact represent harpoons, and if so that is the first sign of this new weapon in Europe, actual examples of which, in bone, do not occur before 14,000 years ago. After that date, however, bone harpoons show up in large numbers at many cave sites in the Dordogne and Gironde regions of southern France where rivers from the Atlantic bear salmon and perch—some of them, as at Mas d'Asil, with a row of a half-dozen barbs on only one side, most, as at La Madeleine and Massat, with mean-looking barbs about an inch long on both sides. There is also evidence that some of the harpoons, in general five or six inches long, were in fact the tines of tridents and forks.

One last weapon of this period, though known so far only in a single example, is what is assuredly said to be a boomerang, made of mammoth tusk, found at Oblazova Rock in Poland and dated to 20,300 years ago. It is twenty-seven inches long, with ridges at one end for a grip, tapered and curved to a blunt point at the other end; it was, like most boomerangs in the world, a non-returning weapon of the hunt to be used primarily on small animals. Modern archeologists have made an exact replica in plastic, with the same weight and density, and have shown it to be accurate even with slight winds at a distance of twenty-five to thirty yards. So far no other similar weapons have been found in Europe, but numerous pictures of hunters using the boomerang turn up in northwestern Australia, at the Bradshaw Rock System, dated to 17,500 years ago, and it becomes a staple of the aborigines there. Archeologists assume a separate origin for the Australian pieces rather than a borrowing, considering how essentially simple and obvious a weapon it is, but for the same reason it is slightly odd that more have not been found in Europe.

It is generally held that Sapiens of this period also used woven nets to trap birds and small game, and though none of the nets survive there is evidence of weaving as old as 26,000 years ago: pottery from Pavlov I in the modern Czech Republic embedded with crossed impressions of woven material, basketry or textiles. Similarly, from the great number of fish bones found in sites after about 20,000 years ago, it is assumed that Sapiens must have had weirs and nets to catch fish, as well as fishhooks (which have been found at roughly 14,000 years ago but were most probably in use long before that). It has also been posited that Sapiens had some kind of portable fencing—this is said to account for the numerous grid-like symbols in so much French cave art, according to the University of New Mexico anthropologist Lawrence Guy Straus—that could be used to enclose and trap a herd of animals driven into the many cul-de-sac valleys of southwestern Europe.

An explosion of weaponry, then, to respond to the crisis of the cold, a challenge to the human role of domination that drew forth an ingenuity and inventiveness not seen before.

And drew forth new hunting strategies as well. It may be too much to say, as Straus does, that "regular, planned, efficient slaughter of large numbers of herd animals and the taking of elusive or dangerous game" *begins* at the time of the glacial maximum, but there is evidence of new and more deadly styles of the hunt, an increase in the number of species hunted, and particularly in the numbers of animals killed, taking human impact on nature to a new level.

The first hunting strategy was based on the new weapons that allowed hunters to attack their prey from a safe distance, particularly the spear thrower and the bow. They could now take on reindeer and giant deer, for example, with much less fear of being gored by antler points, they could dispatch or drive off competitive predators such as hyenas and sabertooth cats, and they could kill the fierce carnivores that they used for fur clothing—wolves, wolverines, and arctic foxes in particular—with much less risk of danger from the pack; being able to attack from twenty yards or more away from such species, on an outcrop above a stream or from behind a protective boulder, gave the hunters a true advantage they had never had before. It also meant that in times of scarcity they could go after a much greater range of small animals that until then had not usually been calorically worth the effort to hunt with just knives and spears, and thus the Sapiens diet in this late glacial period expanded to include rapidly reproducing species like rabbit and hare, birds like partridge, dove, and goose, and an increasing amount of fish, particularly salmon from the Atlantic rivers of southwest France and northern Spain, and from southern Russia.

But however useful these weapons were in taking down single animals, there would of course be more value per effort if it were possible to kill off a large number of animals at a time instead of just one or two. Fortunately for the hunters, the main food species were congregating mammals that would travel in herds and were comparatively easy to divert and drive. Hence other new strategies, as is suggested by the location of the majority of human settlements in the 10,000 years after the glacial maximum: deep valleys like the

Dordogne and the Dnieper, chosen not just for shelter but for certain special physical features they offered — steep cliffs, narrow gorges, dead-end canyons, natural shallow fords — that could be used in planned kills, most often taking one to two hundred bodies at a time in what are called "mass" slaughters.

Using "cliff drives" to force many animals over a precipice was practiced in some places by the Neandertals, and certainly in earlier eras by Sapiens in central and southern Germany and Moravia, but now the evidence for it is far more abundant. Near Solutré, for example, a cave with rich deep-glacial strata in the Massif Central mountains of eastern France, a layer of more than three feet of horse bones was found at the bottom of a steep cliff, representing something between 10,000 and perhaps as many as 100,000 horses, and known with archeological wryness as the Horse Magma. Not far away, along the Vezere River, a series of twenty-one deep pits has been found with many reindeer bones and 20,000-year-old artifacts, a place that the original archeologist called "Les Trappes." And the limestone ridges of the Cantabrian coast are dotted with deep, steep "sinkholes," many leading down into caves, that were ideal for trapping the red deer herds of the area, as attested by what Lawrence Straus calls the "masses of bones" found in cave sites there, particularly at El Juyo and El Castillo near modern-day Santander.

Drives were supplemented by what are technically known as "surrounds," or coralling, by which hunters would descend on migrating herds and force them into narrow, steep-walled, dead-end canyons and valleys from which they could not escape and where they could not even move freely, and so could be slaughtered with comparative ease. Land features that lend themselves to this technique are found throughout southwestern Europe in particular, as around Les Eylies and the Lot valley in southern France and the Basque country of northern Spain, where many cave sites were occupied. The remains of thousands of animals found in the cave of La Riera, one of the westernmost of the Cantabrian sites, testify to the effectiveness of the blind valleys and narrow gorges there as traps for as many as 12,000 years after the glacial maximum. They also testify to one consequence of that effectiveness: bison, which because of their large size were the most desirable of the Cantabrian mammals, predominate in the earliest levels but then are nearly absent for the rest of the time; horses, next, appear for a time and then are absent from later levels; finally there is a concentration on red deer, about a fifth of the meat package of a male bison but the only herd species left in the area. Sapiens could not escape from the reach to total control that always leads to overexploitation.

Still another approach added to the impact of the Sapiens hunters on their surrounding fauna. In the late glacial period more and more settlements were made at or near rivers, both in the West and in the East, and from about 18,000 years ago an increasing number (in France, all large sites) are near shallow fording places at which animals would be forced to cross in their annual migrations. This pattern obviously provided a great opportunity for ambush and mass kills of helpless animals at least twice a year for any herd species of the region—at least until the populations (the animals, that is) were exhausted. Typical of the settlements were four caves (Duruthy, Petit Pastou, Grand Pastou, Dufaure) at the base of the south-facing Pastou Cliff in the far southwestern corner of France, in a narrow valley where two rivers join the Gave d'Oloron on its way down from the Pyrenees to the Atlantic. Because of high cliffs that line that river for much of its length, the best place to cross it is at a year-round ford near the rivers' confluence, just in front of the Pastou Cliff, and it is here that migrating reindeer bands—creatures of habit, by the way—would inevitably pass on their journey up into the mountains every spring and back again to the plains in the fall. From the top of the cliff there is a commanding view of the terrain, and the reindeer herds, visible from afar, could easily be anticipated and attacked by a large band of well-prepared hunters as they crossed the ford; the especially numerous reindeer remains in all these caves attest to the efficacy of the strategy.

It is important to note that these hunters were able to enjoy the fruits of such mass kills only because, thanks to the climate they lived in, they were able to develop systems of storage. In places where the subsoils were permanently frozen, they would have dug deep pits—some stone points have marks that show they were probably used for this—and stored dressed meat for the times in between herd migrations; in other places they could wrap the meat in skins and keep it in the parts of deep caves where temperatures were low year-round. This is new, and portentous: storage means the regular settlement of a single site for a long period, a year or even longer, and the abandonment of transitory sites that would be occupied for only a few weeks or months.

Such a sedentary life style, which is usually thought to have developed only in pre-agricultural societies around 11,000 years ago, would have had the result of moving such people to what the British anthropologist James Woodburn has usefully called "delayed-return systems," in which elaborate dependencies are created, disgruntled groups cannot "hive off" without giving up their food supply, and inequalities and hierarchies tend to develop around the distribution of stored commodities. On the Russian plain in this period, for example, there are a great many places with food-storage pits dug near open-air sites—initially in the middle of a group of dwellings, implying the sharing of stored resources, but later on near only one or two dwellings, protected by walls of mammoth bones probably covered with skins, implying that a favored few were in charge of the meat and probably of its allocation. Inequality seems to have been the price for the storage of these advanced hunters' mass kills.

That Sapiens were now able to use so many strategies for killing, and killing in great numbers regularly through the year, marks still another crucial point in the development of human domination. John Pfeiffer notes that "floods, earthquakes, volcanoes, and other devastations do not operate selectively but wipe out entire populations," and adds that with the mass kill as "an increasingly common phenomenon" people were now able to do the same: "*Homo sapiens* had come a long way in his game drives. He was no longer killing like a normal predator, but like a natural catastrophe, an act of God."

In fact Sapiens had become so good at killing, and violence so much a part of their lives, that they began to practice it on themselves. This should not come as a surprise, of course, because once you have the weapons and skills for killing large mammals, it is not difficult to use them on any one large bipedal animal that may be somehow thwarting your interests in a serious way; it is perhaps more surprising that intraspecies killing does not seem to occur earlier in the fossil record than this time of climatic stress.[1]

There is an earlier suggestion of Sapiens murder, part of a thick wooden pole in the pelvis area of a young man placed in a shallow pit in the "triple burial" at Dolní Vestonice, near Brno, 27,640 years ago. I say "suggestion" because it is not at all clear that this piece of wood was actually part of a spear thrust into the man while alive, and it may well have been, like nearly two dozen other pieces of wood found on and near the other bodies, the residue of a funereal fire of branches that was extinguished before it could damage the bones of the interred. Similarly, a projectile point found embedded in the spinal column of a child at the Grimaldi caves in Italy and dated to 25–18,000 years ago is less likely to have been evidence of murder than of accident, since children are not usual targets of murder.

The first unequivocal evidence of human murder comes not from Europe but from the Nile Valley, at Wadi Kubbaniya near the Egyptian-Sudanese border, where the dry and windy conditions of the glacial maximum would have made water a precious commodity and the Nile a favored settlement area. There an American archeologist, Fred Wendorf, then of Southern Methodist University, uncovered a young man who was buried about 20,000 years ago, with three wounds: a broken left forearm, presumably from warding off a blow, a spear point in that upper arm surrounded by partly healed bone, and two spear points embedded in his pelvic bones. Wendorf believes that people were regularly fighting over the best watering spots in the valley, where game and waterfowl would come in the winter and fish were plentiful in the spring spawning season, and that this young man had gone through three separate scraps over the course of several months. "The last one" in the pelvis, Wendorf says, "got him."

There is no evidence for regular fighting, or warfare, at this point, however, and to assume that it occurred would be a considerable leap based on the bones of a single murdered man, even if he was in three fights. It is perhaps better to posit that the strains of the period might lead to murder within a tribe whose bands could no longer so easily hive off into unoccupied areas when tensions between them mounted; or the killing of a member of another tribe that was encroaching upon a staked-out territory and its supply of water or game; or the sanctioned removal of one band member who had somehow transgressed the increasingly complex rules and rituals of the group and was disturbing the fragile harmony necessary to get through difficult times.

That, at any rate, is how archeologists generally interpret the various speared human figures that show up for the first time in the cave art of Europe about 20–18,000 years ago. At Cougnac Cave, for example, in the Dordogne region, one set of figures includes a man pierced by at least six lines, presumably spears, and another shows just the lower half of a human figure pierced by lines in the lower back, rump, and thigh. Just twenty-five miles away, at Peche Merle, there is a quite similar painting in red of a standing man (a small line at the thigh indicating his sex) with four lines on each side as if he were being pierced by spears; not far from that, at Sous-Grand-Lac, is another standing man (this time with an erection to indicate gender) being pierced in the back or chest by three lines, two of which have barbs indicating that they are harpoons. Several other Dordogne caves (Lascaux, Gabillou) have slightly more ambiguous figures that might represent killings, but to show that this theme was not limited to that region, at Cosquer Cave some two hundred miles away on the Mediterranean there is a striking engraving of "the Killed Man" from 18,500 years ago, about eleven inches long, at a prominent place in the entrance chamber. It shows a figure lying on its back (there are in fact no indications of gender), with an arm and legs extended upward, a lightly etched line running through its chest and a long, very heavily engraved line — a spear or harpoon — that appears to hit the back, pass through the body, and penetrate clear through the skull, a most convincing picture of homicide.

There is no longer any question among archeologists that these figures represent people murdered at the hands of people, but there remains some question of *why* they were made — whether to commemorate some event, an execution or a sacrifice, or to symbolically exorcise some evil spirit, or to cause injury or death by imitative magic. There are adherents for each position, but without some further evidence there is no good way to know the reasons and the rituals that went with this kind of art. All one can say, along with the French prehistoric antiquities experts Jean Clottes and Jean Courtin, is that "the Killed Man of the Cosquer cave tells us one thing certainly, that in this Paleolithic world so often painted in idyllic colors, the idea of murder or of execution, of whomever for whatever reason, was a familiar one."

Once the idea of individual killing becomes "familiar," the idea of *group* killing should not be far away. While there is no evidence for band- or tribe-level "warfare" in Europe until much later — the six cases of humans with bones pierced by points in the period 14–10,000 years ago are at separate sites — it would not be much of a stretch to imagine intertribal rivalries for game in times of scarcity, and those rivalries occasionally breaking out in armed confrontations. Indeed, the Harvard biologist Edward O. Wilson has asserted that "intertribal aggression, escalating in some cultures to limited warfare, is common enough to be regarded as a general characteristic of hunter-gatherer social behavior," and a recent cross-cultural study of hunter-gatherer societies in historical times suggests that "those societies which rely heavily upon hunting are more likely to engage in frequent warfare, and further that if large animals are hunted, the likelihood of frequent war increases"—and that certainly describes the societies of deep-glacial Europe.

And we do know that Sapiens societies were capable of violent conflict on a group scale, from the site that is generally accepted as the first evidence for "warfare," Cemetery 117 near Jebel Sahaba, northern Sudan, dated to about 14–12,000 years ago. Here were found fifty-nine well-preserved skeletons, carefully buried, of whom twenty-four, men and women, showed evidence of being hit by arrowheads (or small spear points), most in the chest and back, others in the lower abdomen, and a few in the skull through the lower jaw, as if they were shot or skewered while on their backs; one skeleton of a young woman had twenty-one stone points in her body, while another of a young man had evidence of nineteen wounds, suggesting some sort of ritual killing. The excavator, again Fred Wendorf of SMU, acknowledges that the cemetery was used for several generations, but he believes that these skeletons were buried at the same time and thus show signs of being casualties, not so much of war as we think of it traditionally between men, but perhaps of a fierce raid or ambush in which women would be victims as much as men. Either way, human violence against humans, at a period as close in time to us today as to the people buried at Sungir.

And then the animals started to die out.

The harsh climate surely took its toll on a number of warm-adapted species, but it is highly probable that human hunting was ultimately responsible for the remarkable number of species that were made extinct in Europe in the period between 20,000 and 10,000 years ago. At least twenty genera of mammals—genera include a variety of species—died out or disappeared regionally at this time, including all the large grass-eating species and a number of carnivores, something on the order of a devastating *60 per cent of the large mammalian population*; nothing like this rate of extinction had ever been experienced before in Europe.[2]

Take the emblem species of this era, the woolly mammoth. It was amply distributed in much of Europe after the glacial maximum, but its numbers dwindled in western Europe and it was extinct there by around 18,000 years ago — once common in cave art (as at Chauvet Cave, 35,000 years ago), it vanishes about that time (none are among the eleven species at Cosquer Cave, 18,500 years ago) — just during the period when food resources were in shortest supply and hunting most intense. It became rare in the rest of Europe after about 13,000 years ago, even on the Russian Plain where it had thrived earlier, and the last individuals succumbed around 12,200, at Praz Rodent, Switzerland. It may have lasted longer in Siberia, but there too it disappears from the record by around 10,000 years ago, except for one remnant population on an Arctic island that seems to have escaped predation until 4,000 years ago.

What happened? The climate was difficult, no doubt, in the northern regions where grasses gradually died out, and some mammoths might have perished before they migrated south to the refuges of the Russian Plain; then too, in the millennia after 18,000 years ago there were quite a number of warmer climatic oscillations, and some periods when forests grew back in places to displace the mammoths' favored grasslands. But mammoths, a species probably 400,000 years old, had survived such periods in the past, both the deep glacial colds and the intermittent warms, as well as the fluctuations in between: the only thing different now was that human hunters were out to harvest them with potent new weaponry and scarcity-prodded ingenuity. As the South African anthropologist Norman Owen-Smith has pointed out, given their slow reproductive rate — a gestation period of about three years, it is thought, similar to that of African elephants today — "there's no need to consider any other explanation beyond direct human predation as a cause of those extinctions."

Hunting we can understand. But hunting to extinction? That would obviously seem to be a not very successful long-term strategy for a food supply. Why would hunters eliminate a species that they depended on?

The short answer is that they probably did not know what they were doing until it was too late. They took a few animals regularly, maybe drove a herd or two over a cliff, and had no idea what the eventual effect would be on the local population: there had always *been* mammoths, why would there not always *be* mammoths? Besides, if one tribe refrained from killing them, even if it realized that the mammoth numbers were dwindling, how would it know if another tribe nearby would make the same decision? And if the meat supply from other species was scarce, there may have been no good alternative to killing mammoths, who were after all a very economical source of meat: an adult male would supply something like 2,500 pounds of meat, which would feed a band of twenty-five people (again, at 4.4 pounds a person) for three weeks if it were cured or stored. It would require a lot of rabbits and partridge to equal that, and a lot more time and energy than killing a single mammoth.

What's more, according to an elaborate simulation of human population density and numbers of mammoths run by Steven Mithen of the University of Reading, there would not need to be any great rampage of hunting to wipe out a mammoth herd. If mammoths were like modern elephant populations it would take as many as twenty years for them to reproduce their full number, so a regional extinction could happen in a fairly short period if only a score of adults were killed a year. Mithen figures that "the required hunting intensity to push mammoths to extinction in all possible combinations of mammoth/human population densities and population characteristics is only 0.75 mammoths/person/year"—or nineteen a year for a band of twenty-five people, roughly one every three weeks, just the right rate for a regular supply of meat. It does not take much imagination to see how a hunting society, particularly if it was skilled at taking adult animals, could drive a species to regional extinction in a very short order.

Extinction. That is a heavy charge to make. It proclaims that Sapiens had become a species so technologically powerful, so effectively deadly—and so psychologically fixed on its superiority to the rest of life—that it could eliminate one whole other species from its habitat. Not necessarily by intention—in fact, probably by accident, not taking the time and care to figure out the long-term effects of its actions, though surely as hunters intimate with the ways of their prey they knew that mammoths took a long time to gestate and a long time to grow to reproductive age, and had arrogance not overpowered humility they might have realized their effect on the dwindling herds and switched to other animals.

But regional extinction was not limited to mammoths, suggesting that something more than ignorance was at work. Elephants were eliminated in Europe by about 14,000 years ago, the woolly rhinoceros by 11,000, the giant deer at the same time; in fact by 10,000 years ago sixteen out of twenty-seven large mammalian species (technically, those above eighty-eight pounds) were gone from Europe, never to return. The marked climate change after about 11,000 years ago, as the glaciers retreated and the Ice Age came to an end, has been cited as a cause for these extinctions, but again, all these species had experienced periods of similar weather many times in the last half-million years and had no reason to succumb to this one. It well may be that they were not all hunted to extinction, though some that were, like the giant deer, horse, mammoth, and saiga antelope, were prime meat sources; conceivably the elimination of just those four species caused so much disruption of the ecosystems, severely changing the nature of the grasslands they had been feeding on or depriving other predators like hyenas and sabertooths of their regular meals, that other species could not survive.

Nonetheless, the human hand seems primarily culpable here, and the conclusion is almost irresistible that the frenzied intensification of hunting in the period after the glacial maximum, reflected also in the immense number of animal cave paintings of the time (at least 80 per cent of the paintings were done in the 8,000 years after 18,000 years ago), was so powerful and encompassing that it drove hunting bands to the excess of extinctions. Besides—and this is a terrible thought that seems all too likely—it may not have mattered to them: there were still herds of reindeer, elk, bison, auroch, red deer, and fallow deer to kill, and nature was always there to provide more food for human hunger, if the right ceremonies and rituals of control and replenishment were followed.

And extinction is clearly the deadly pattern of humans elsewhere in the world. We have already seen the result of Sapiens encroachment into Australia, where an incredible twenty-three of twenty-four large mammalian genera, containing fifty-five species—mostly slow-moving animals, with no experience of humans and no instinct of flight from them—were driven to extinction, starting around 46,000 years ago. Again, hunting seems to have been the principal agency of destruction, though climate variations and such human practices as deliberate fire setting (attested by archeological charcoal deposits) may have played a part in altering the vegetation and thus disrupting the herbivores' diets and ranges. Some paleologists have posited that climate change, particularly aridity after 26,000 years ago, was the leading reason for extinctions, and point to the scarcity of "kill sites" in Australia with both large numbers of bones and human artifacts. But Paul Martin, the University of Arizona geoscientist who champions the idea of humans as the primary cause of late-glacial extinctions, has argued that the sites are so few because the movement of people across the continent was so fast and no particular site was settled for long, archeologically speaking. And he points out that before the human there was only one swift predator, a giant "killer possum," and that the mammals there evolved generally without speed as a defense mechanism. He concludes that conditions were especially ripe for predatory extinctions: "The arrival of a potent and deadly species, *Homo sapiens*, landing on a continent that had previously known few large cursorial carnivores . . . seems uniquely favorable for overkill."

The case is even starker in the Americas. The dates for the first Sapiens migrations from Asia into the New World are not absolutely sure, but it is generally accepted that they could not have been significant until after 14,000 years ago. It is at that time, when there was still a frozen land bridge across the Bering Strait connecting Siberia to Alaska (ice sheets locked up so much water that the seas had fallen by up to 460 feet), that a gradually warmer period would have allowed forests to reappear in the sheltered valleys along this route, providing fuel and shelter for human bands, and for some species of large mammals that also began their migrations out of Asia. The earliest sites in Alaska (Broken Mammoth, Ushki Lake, Nenana Valley, Healy Lake) date to between 14,000 and 12,000 years ago, but it is possible that people could have made their way south earlier than that as the huge ice caps covering most of Canada began to melt and a narrow corridor opened up just east of the Rocky Mountains; one site near Edmonton, at the southern end of this corridor, is dated to roughly 12,000 years ago, and there are even posited (and controversial) dates of 12,500 years ago for two sites thousands of miles away, in Meadowcroft, Pennsylvania, and Monte Verde, Chile. (There are even claims for human sites of 14,100 years ago for Pikimachay Cave, Peru, and 13,000 years at Taima-Taima, Brazil, but if accurate that would suggest some kind of coastal rather than overland route because the ice sheets would have been impenetrable then.) Whenever the journey began, it is certain that by 11,200 years ago humans had penetrated as far south as New Mexico—and, in a remarkably swift passage as these things go, had reached the very tip of South America by 10,500 years ago.[3]

One reason for this rapid advance of human populations over the Americas may have been that they were simply wiping out most of the large hunted animals as they went, regularly decimating their local resource base in a matter of years, transforming the habitat, and were forced to move to new territories with new prey. These were very expert and cunning hunters, mind, almost certainly with all the weapons known to their European counterparts, and most of the animals they encountered—mammoths, mastodons, camels, horses, yaks, ground sloths, llamas, deer—were native to the continents and had no experience of human hunters nor the instinct of white-eyed fear that had evolved into their African and Asian equivalents. Some thirty-five genera died out in North America by about 11,000 years ago, 73 per cent of the larger animals and all of the biggest ones, a number that may be as high as 50 to 100 million animals; as many as forty-six genera became extinct in South America—were the hunters getting more proficient as they moved along?—and an astonishing 80 per cent of the large mammal genera.[4] Such high numbers imply that not all the species were hunted to extinction—four species of ground sloths, for example, not likely to be highly prized meat, die out by 10,000 years ago—and again, that alteration of habitat, eliminating key species and using fire for game drives, was responsible for much of the devastation. Those species that survived this onslaught in the Americas, such as reindeer, bison, and moose, were all mammals that had originated in Asia and spent long millennia with human hunters, and though they all were hunted—and became staples when the other species became extinct—they were wily enough not to be driven to extirpation.

Some paleoanthropologists still cling to the notion that humans could not have caused such widespread and near total slaughter, and theorize that the extinctions must have been due to human-induced diseases for which the American animals had no immunity, or severe climate change after the glacial epoch, or human disruption of ecosystems through fire or the elimination of the big browsing mammals. There is even one theory that what the Sapiens hunters really did was kill their fellow carnivorous species first—bears, sabertooth and scimitartooth cats, cheetahs—because they were dangerous animals and competitors for the large game, and that with the loss of these "key species" the population of the herbivores exploded, grasses and trees were overgrazed, large sections became uninhabitable, and the big mammals died out by themselves without human agency. There is just not enough substantial evidence to sustain any of these lines, though the battles around them have been long and often bitter, and there are still some who keep searching for signs of climate change as the culprit even after it has become clear that the climate of this period was in no way extraordinary, that it changed gradually overall after the Ice Age, and that most of the extinctions occurred before it was over. The large majority of paleoscientists now would agree with Richard Klein's simple assessment that "what differentiated the end of the Last Glaciation most clearly was the presence of more advanced hunter-gatherers." More advanced, and more deadly—and more steeped in the gore of conquest.

In the end, after some 8,000 years—by roughly 10,000 years ago—100 per cent of the animals over 2,200 pounds, the large mammals in the mammoth family (*proboscidea*), were extinct in Eurasia and the Americas, as were 76 per cent of the mid-sized animals weighing 220 to 2,200 pounds, and 41 per cent of smaller ones between about 10 and 220 pounds. That is a record of almost unbelievable enormity: "a general disaster" for the other species of the world, the historian Alfred Crosby has said, "nothing more spectacularly devastating in tens of millions of years" of the earth's history. Richard Klein adds that this phase of our species' evolution marked the transformation of humanity "from a relatively rare and insignificant member of the large mammal fauna to a geologic force with the power to impoverish nature." What greater condemnation of a way of life could be imagined?

One final note. As if to confirm our role in animal extinctions, the pattern of rather sudden die-off after initial human intervention was repeated again and again in the millennia after 10,000 years ago, in New Zealand (twenty-eight species extinct within a few centuries), the Bismark Archipelago, Tahiti, Fiji, Tonga, New Caledonia, Marquesas, Chatham Islands, Cook Islands, Solomon Islands, Hawaii, Cyprus, the West Indies, and Madagascar. No wonder Edward Wilson, with only slight exaggeration, has called Sapiens "the serial killer of the biosphere" and its habitats "a slaughterhouse."

A crucial question arises: How, if humans managed to kill off so many species, including a great many of the larger ones representing the most efficient packages of meat, and if in the process they disrupted many of the environmental niches where the animals had been, could they find enough food for survival?

One answer is that there were enough surviving species in most places for humans to prey upon, and humans had learned by now how to survive almost anywhere and in almost any conditions by a selective and adaptive exploitation of whatever species nature left them with. The species now were often smaller and harder to catch, it is true, less desirable (and less prestigious—no bringing home mammoth tusks), and required much more work per calorie supplied (as with fish and shellfish, for example). But these animals had been part of the intensified diet for millennia and it could not have been difficult to concentrate on them now.

And humans were helped exceedingly by the happenstance that from about 16,000 years ago the world's climate had begun its long climb out of the Ice Age, punctuated by only a few short periods of returning cold, so that by 11,000 years ago, when the big animals were mostly gone, there began a long period of generally benign temperatures and nurturant precipitation. In Europe, for example, the retreat of the glacial sheets allowed the return of deciduous and conifer forests where there had been only frozen tundra and polar deserts, and grasslands where there had been only barrens, along with the return of some species adapted to the woodlands and the rivers swollen by glacial meltwaters. These would include small and midsize animals (aurochs, several species of deer, pigs), marine and riverine species (fish, especially salmon, and shellfish), birds (ducks, swans, herons, cormorants), and of course many warm-weather plants that had been unavailable in the deep-cold era.

The extent to which all these species were used is difficult to determine, especially for floral ones that do not easily leave traces, but the evidence is strong that the Sapiens populations practically everywhere, and especially in the later stages of this period, not only survived but thrived. Almost everywhere there was an increase in the number of known sites and their density on the land, suggesting that populations were swelling and desirable settlement areas were filling up. In Europe, with its lengthy coastlines, land that had been successfully settled for many millennia was increasingly lost to the rising oceans, and populations were forced to move to the interior: the area of southern Germany started to be resettled after about 15,000 years ago, similarly northern France, and then Denmark and even Britain; populations moved out of the Russian plain both east as far as Moravia along the Danube and northwest into Russia and Kazakhstan. In Africa, after about 14,000 years ago, there was a similar increase in population in the Nile Valley, the northwest coast, and sections of western, eastern, and southern Africa, attested to, especially in the lower Nile, by what the archeologist Mark Nathan Cohen calls "a striking number of indices of high population density and significant population pressure," as well as ample evidence everywhere of an increasing number of small animals (such as lizards, tortoises, and rodents in the Maghreb) and marine fish and shellfish in the diet. And in the Levant, and northward into the Taurus Mountains of southern Turkey and westward to the Zagros Mountains of western Iran, again the number of sites, particularly open-air sites, increased dramatically after 13,000 years ago or so, especially along the Mediterranean coast and in previously unoccupied areas in the hinterlands like the Judean desert and the Jordan Valley.

The rise in human numbers—by one estimate as many as 16,000 people were in the Levant and Mesopotamia by 10,000 years ago and perhaps 45,000 a millennium later—obviously increased the competition for edible species and repeatedly called forth new methods of gaining food. One obvious resource was the increasingly abundant plant world, nurtured by the warmer and wetter climate, plus—perhaps most important—by an increase of 25 per cent in atmospheric carbon dioxide from 15,000 to 12,000 years ago, which is estimated to have increased plant productivity by up to 50 per cent. So in addition to the nuts and fruits and berries that came with the returning forests, there were several favored places around the world with moderate climates and long growing seasons where there were numerous stands of large-seed cereal-grass species like wheat and barley as well as pulses like peas and beans.

It is impossible to say where humans first got the idea of cutting off the seeds of growing grasses, threshing them, grinding them into an edible paste —and then adding water for a porridge or heating them into a loaf—but archeological evidence shows that the practice could have been started as early as 15,000 years ago and was fairly widespread by 11,000 years ago. One indication comes from flint blades with a distinct shiny surface along the cutting edge that archeologists call "sickle sheen," made when they are used to cut abrasive cereal grasses. They are found in the Levant at two sites of about 15,000 years ago (Ein Guev and Nahal Oren, in Israel) and are common there after 11,000 years, at many sites in the lower Nile valley from about 14,000 years ago on (including Tushkla, near Abu Simbel, and Kom Ombo, near Aswan), in the Maghreb at about 12,000 years ago, and in Europe at between 12,000 and 11,000 years ago (for example at Duruthy, a cave in southwestern France, where half of the over 1,500 flint blades have sickle sheen). Another set of certain indicators is the large number of grindstones and pestles used for plant processing that are found as early as about 14,000 years ago in the Nile valley and become common in the Maghreb after 12,000 years, in the Levant after 11,000 years, in southern Africa and Mexico after 10,000 years, and in Japan after 9,000 years.

This widespread harvesting of wild grasses is somewhat peculiar, though, since although they are easy enough to gather by cutting the seeded tops into a basket or blanket (and a year's supply can be gathered by a small family in three weeks), grasses must be harvested during a short season in spring before the seed shatters and drops to the ground, take a lot of work to process before they are edible, provide basically uninteresting fare no matter how prepared, and are poor sources of protein compared to meat or even nuts. Why then did they begin to play such an important part in the human diet after about 13,000 years ago? It is true that in many places grasses were growing in profusion in dense and extensive stands, enough to sustain fairly large populations, that they are easily storable in underground bins and can last a year if kept dry, and that their nutrients are readily absorbed and stored in the human body. But given their drawbacks—and that humans had never bothered with them before—it is likely that grasses were made use of by necessity and not by choice at a critical point when most other sources, animal and vegetal, had been overexploited and human numbers exceeded the immediate carrying capacity of the land. Mark Cohen, in his *The Food Crisis in Prehistory*, sees it as "very unlikely that a human population would have settled down near a field of wild wheat for the dubious pleasure of harvesting, threshing, and grinding the grain and eating the gruel year-round. . . . Grass use and seed storage thus probably emerged as behaviors practiced by people under population pressure when the need for stored calories outweighed the costs involved."

One of those costs, probably never anticipated, and one that was to have wide-ranging social effects over the next millennia, was the need to set up camp not far from the stands of wild grasses—*and* to stay there year-round to be close to (and protect) both the grasses and the stored-up seeds of the springtime harvest. For populations that over the millennia had been used to moving around a range of camps during the year, sometimes covering hundreds of square miles, this sedentary life must have created a whole new set of problems, not least the difficulty for one family or band of "fissioning off" and leaving the village to settle elsewhere when social tensions and frictions arose, as had been done regularly by hunter-gatherers for eons in the past.

We can best see the process of wild-grass harvesting leading to sedentary societies in the aptly named Fertile Crescent, an area of the world that had a climate so ideal for floral life—long dry summers and mild wet winters— that by 13,000 years ago it had more than a hundred edible plants, thirty-two of the fifty-six most beneficial large-seeded wild grasses, and abundant peas, chickpeas, and lentils. This area, stretching in a semicircle from the southern Levant up to southern Turkey and around to the Tigris and Euphrates valleys, was remarkably rich in soils, with a high diversity of species, including by far the greatest number of annual self-propagating ones that are best as food sources. (Annuals, unlike perennials, have to put their energy into seeds, not stems.) It was along the Mediterranean coast here, where some marine foods were available and not far from large stands of emmer and einkorn wheat, barley, and flax, that a culture called the Natufian flourished from roughly 13,000 to 10,300 years ago, settling in permanent and semipermanent villages, a number of them with stone- or clay-walled buildings (as at Ain Mallaha in Palestine and Tell Mureybat in Syria), and many with evidence of underground storage pits (and of higher numbers of hanger-on species like rats, mice, and sparrows than at hunter-gatherer sites). It is possible that these people still hunted for some of their food, since gazelles are represented on a number of figurines and decorated tools, the first flowering of art in the Levant, but that species was eventually overhunted and its numbers diminished in Natufian times, and hyena and rhino were extinct. Most of the artifacts point to dependence on wild cereals and pulses: sickles in great profusion (as at Kebara Cave and El Wad), storage pits, mortars (from Nahal Oren, El Khiam, Kebara, and elsewhere), grinding slabs (as at Ain Mallaha and Zawi Chemi) said to be "quite numerous" after 11,000 years ago, and bowls in large numbers at almost every site. As Ofer Bar-Yosef, the Levantine expert, sums up the case: "Natufian communities practiced intensive and extensive harvesting of wild cereals."

In the middle of the Natufian period in the Levant came another drastic change in climate that might have been the catalyst for the passage from wild-grass foraging to outright wild-grass domestication and the beginning of agriculture. Over a period of about a thousand years, from roughly 12,500 to 11,500 years ago, rainfall diminished and temperatures dropped sharply again—not as drastically as during the severe cold of the glacial maximum but with enough severity to change the patterns of plant life in favor of cold-tolerating species and to begin diminishing the range and number of wild cereal grasses. It is probable that in many areas people either continued or reverted to hunting as some cold-adapted prey species returned, but in those regions like the Levant where animal populations were limited and cereal grasses and pulses had come to play a central part in the diet, some other solution was needed; especially so if the population density of that region meant that there were no other convenient places to move to. So it would have made sense at this time, for people who knew enough about the lives of these plants after millennia of harvesting them, to come up with the idea of growing the cereals from seed, in the most favorable warmer areas, and watering them independently without counting on the now-uncertain winter rains. It was really not such a great step to take for a species long used to manipulating nature and attempting to control its resources, and the food crisis would have provided the impetus. Of course it would have consequences—indeed, it led to the greatest change in human society since the advent of large-animal hunting itself—but the need was great and the Sapiens species had relied on its capacity for ingenuity and adaptability for many thousands of years.

This is clearly what happened in the Fertile Crescent. At least fifty sites in that region show that sometime after 12,000 years ago people began domesticating the grasses and pulses that had once grown wild and began farming as their primary source of food. One scenario recently put forth by a team of European scientists suggests that it was sometime around 11,000 years ago, when the fields of wheat and barley were likely diminishing, that some people in southern Turkey, near Karacadag Mountain on the upper reaches of the Euphrates, began to plant and tend seeds of wild einkorn wheat. Modern research has shown that certain cultivated einkorn plants are remarkably similar in genetic makeup to wild einkorn still growing near Karacadag Mountain, which therefore, according to Manfred Heun of the Agricultural University of Norway, is "very probably the site of einkorn domestication." Other sites offered as the origins of agriculture are the corridor from the Dead Sea to Damascus at 10,000 years ago; the village of Abu Hureyra in northern Syria on the upper Euphrates around 9,600 years ago (or even earlier); and the Jordan Valley, where emmer wheat and barley (which have the biggest and second-biggest cereal seeds) were grown about 9,000 years ago.

But wherever agriculture began, it was somewhere in the Fertile Crescent and sometime after 12,000 years ago, and it was firmly fixed there throughout the whole region by 8,500 years. Remarkably, and still largely inexplicably, it was also begun in many other areas of the world not long after it was established there: in northern China, where technological changes indicate intensive plant procurement from about 11,600 years ago; in the Andes, where squash and gourds were grown from 11,300 years ago (and lima beans from about 8,000); in central Mexico, where squash was cultivated some 10,000 years ago; in New Guinea, possibly at 10,000 and certainly by 7,000 years ago; in southern China along the Yangtse River, about 8,000 years ago; in Egypt, presumably derived from the Levant by 7,000 years ago; in west Africa, from about 4,700 years ago; and in the Mississippi watershed, around 5,500 years ago. For such an essentially odd and basically onerous system as seed domestication to have arisen independently in these eight places over a comparatively short period of 6,000 years seems to beggar explanation, but there is no evidence of any kind that this practice was spread by dispersal (of knowledge or people) from the Fertile Crescent, and we must fall back on the idea that climate crises similar to that in the Crescent, in places where animal species were diminished but plant species were familiar, evoked similar responses in similar kinds of people.

In any case, in these eight favored areas—and eventually by dispersion into Europe and India, eastern and southern Africa, the rest of North and South America, and Japan—agriculture became the established way of life for the great majority of the world's people. And when I say "way of life" I mean that in the fullest sense. Agriculture was not simply a way of getting food, satisfying one basic human need; it shaped and regulated every element of the culture every bit as much as hunting had done, affecting plant and animal ecosystems, settlement patterns, family relations, mating choices, systems of governance, and much else besides. Above all it cemented in the human mind the psychology by which people understood their world: it was *we* who chose what seeds to plant and where, what forests to cut down and fields to fire for them, what would be declared weeds to pull, what waters to divert to irrigate them, in short what species were to live and die, and when and how. Agriculture was a superb demonstration that *humans could control nature* (or believe they could), could literally domesticate it and place it under regular and systematic human will and design: "We began to imagine ourselves," says the archeologist Gordon Hillman of University College, London, an expert on agricultural origins, "masters of the environment." Or as Niles Eldredge of the American Museum of Natural History has put it:

> We removed ourselves from the fundamental position in nature that we had
> heretofore shared with absolutely all other species since life began: we abruptly

stepped out of the local ecosystem. We told Mother Nature we didn't need her anymore; that we could take care of ourselves. . . .

What does it mean to live *outside* ecosystems? It means that our interests no longer dovetail with those of the natural world around us. . . . Inventing agriculture in a very real sense was tantamount to declaring war on local ecosystems.

Hunting had certainly had its impact on local ecosystems, especially hunting to extinction, but for the most part and the longest time it was no more harmful to nature as a whole than any other system of predation. Now, with deforestation, dams and irrigation, soil exhaustion, extensive settlements, and all that goes with agriculture, almost all natural systems were disrupted and degraded. We were declaring war not just on a species but on a world.[5]

That terrible and portentous attitude was surely behind the thinking that led to the next round of domestication: of fellow animals. Again, this seems to have begun in the Fertile Crescent, which had four of the easiest mammalian species to domesticate—goat, pig, sheep, and cow—out of the only 14 species that have ever allowed themselves to fall under human control. (It turns out that for domestication an animal must live in herds, with follow-the-leader hierarchical systems, be amenable to fencing, have a placid disposition, be able to eat the foods that humans could supply, have fairly short growth and birthing periods, and be able to breed in captivity—criteria that 134 of the 148 large mammalian species never satisfied.) Whatever possessed humans to think of and carry through such a process is lost to the prehistorical mists, but we can assume that once they had fenced in wheatfields for convenient food (and protection from being eaten by other species) it was not that much of a leap to try fencing in animals (or at least controlling herds) for convenient food. One study indicates that goats and sheep were domesticated at Abu Hureyra in northern Syria as early as 10,000 years ago (and a mitochondrial DNA study of goats indicates that the first domestication can be dated to just that time), another locates goat domestication at Ganj Dareh in the Keremanshah Valley of Iran at the same time and at Ali Kosh in the Tigris Valley of Iraq maybe a thousand years later, and another locates sheep, goat, and pig domestication at Umm Qseir in northeastern Syria between 8,000 and 7,000 years ago.

Bone remains from sites such as these, replicated at many later sites, indicate that the initial strategy of these people was to use the animals primarily as a meat source, culling the younger males at an early age (generally one to two and a half years for sheep and goats) in the second half of the year when spring crops were diminishing, and leaving the older males and females to produce new annual provender. (It is possible, as some have argued, that the animals, particularly sheep and goats, were also used for religious sacrifices, as in historical times.) Only later was it gradually discovered that they could also be used for (or bred to supply) milk, wool, fertilizer, traction, and transport — and, sometime around 5,000 years ago, when agriculture had displaced most of the ancient hardwood forests, for fuel from their droppings. There were only four domesticated species — cows were added around 8,000 years ago — but they were used to the fullest.

I mentioned consequences. The domestication of plants and animals — "enslavement" might be a more appropriate word to give the right sense of the new relationships — produced pretty much the same effects in societies throughout the world, and these were mostly deleterious, which is why a sober academic like Jared Diamond, a physiologist at the UCLA School of Medicine who has studied it extensively, could call it "the worst mistake in the history of the human race." One might argue that the "mistake" was forced upon some people where game animals were scarce and vegetation flourished, but the solution at all other times in human evolution had been migration to regions where animals were more numerous, not the aggressive cultivation and planting of various grasses and grains: that was new, and fateful.

Perhaps the first important consequence was an increase in population numbers and densities. Farming and herding allow a significantly higher yield per acre of land than hunting and foraging, and wheat and barley in particular are highly productive, so larger populations could be supported — a hundred times greater than in hunting societies — and larger populations are what farmers always want anyway, given the laboriousness of their jobs. This population growth was apparently achieved in most places by having women give birth to more children during their reproductive years, the birth intervals being much shorter for farm families that can wean infants on to milk and gruel and not have to extend female lactation (and hence infertility) for years as hunting families must; it is also probable that wild vegetal abortifacients used earlier to keep populations low were displaced by ever-growing farms and fields. In a very short time clan sites became villages, villages proliferated everywhere, and some of them grew into small, densely packed cities: Jericho, on the western edge of the Jordan Valley, grew from about fifty or a hundred people in Natufian times to perhaps a thousand (some estimates double that) by 10,000 years ago; Catal Huyuk, in Anatolia, grew to a city—the first real city—of perhaps five thousand people around 8,000 years ago, probably the upper limit to which any strictly agricultural society could grow.

But think of what this means. People in sedentary communities of over fifty people are living as no one had ever lived before: they would need to create all sorts of new political, economic, cultural, and social institutions and policies to handle complexities at those scales. It would not have been possible, for example, to maintain communal farms and herds at that scale, and so farm plots and herds would become private property belonging to a family or extended family, from which others would be excluded. Individuals would *own* the earth, lay claim to some part of nature as rightfully theirs, on which would be tended seeds that they owned and animals that they owned, a whole set of concatenations that had played no part in human life before. Gone the ancient rules of reciprocity and sharing, for no goat shepherd is going to give his hungry neighbors one of *his* animals for dinner, as was regularly done in hunting bands when the men came in with a catch, or he would soon be impoverished. Gone too the life without possessions (or limited possessions) imposed on mobile hunting societies, for now with a sedentary population one could have all the possessions that one could accumulate, from goats to grindstones and animals to acres, and the more the better. This is a complete and drastic alteration of the customs by which people had lived for millions of years, one that only accelerated and strengthened the impact of the human presence on the earth and allowed its capacity for domination to be fueled by its newfound evolved capacity for covetousness and self-interest.

Population accumulations and densities had other consequences. Diseases, many from the domesticated animal populations now living in close proximity to humans for the first time, had fertile territory to spread in, and communicable "crowd diseases" that would have died out in the small populations of hunter-gatherers instead flourished (measles, smallpox, and tuberculosis from cattle, flu and pertussis from pigs, plus plague and cholera). Parasitic diseases took hold in larger settlements with water contaminated by human feces, while insects attracted to stationary animals and breeding in standing irrigation water brought yellow fever, trypanosomiasis, dengue, and (at 8,000 years ago in Africa, 4,000 in the Mediterranean) malaria. All in all, a whole panoply of maladies arose that had never bothered human populations seriously before, and it is little wonder that human life spans quite rapidly grew shorter—hunter-gatherer women generally reached the age of forty, sometimes fifty, men fifty or even sixty, but agriculturalist longevity was in general ten years shorter.

Another apparent effect of crowding was a decrease in body size, because when populations expanded to the limit of their food growing, as farming settlements inevitably do, then, as Christopher Stringer of the British Museum explains, "humans therefore had to drop in either number or size, and evolved the latter course." Average body height of the Sapiens hunters was about 6 feet, of the women maybe 5 feet 5 inches, but as early as 5,000 years ago the average height of agriculturalists was 5 feet 3 inches for men and 5 feet for women. (Today in the West the average male height, of only 5 feet 8 inches, still has not returned to the stature of the hunters.) Even more alarming was a decrease in brain size, of 8 to 10 percent, after the beginning of agriculture, perhaps as a result of the tediousness and repetitiveness of farming and herding as well as in response to the social overload of larger settlements, and possibly also because the continual concentration and information processing of the hunter in the wild was no longer necessary. It is as if, John Allman says in *Evolving Brains*, having domesticated plants and animals, "humans, through the invention of agriculture and other cultural means for reducing the hazards of existence, have domesticated themselves"—and, like domesticated animals themselves, grown smaller in the process.

The generally inferior quality of the food that the early agriculturalists ate also probably contributed to these shrinkages. Wheat and barley are higher in protein than rice and corn but are still not particularly nutritious, are deficient in amino acids and vitamins, and are high in carbohydrates, supplying bulk but not valuable calories; herded animals could supply protein, but a family would have to be careful to limit its culling lest it interfered with their propagation. One study of early farmers in North America has found that compared to their hunter-gatherer predecessors, they had a nearly 50 per cent increase in tooth-enamel defects, indicative of poor nutrition, and a 400 per cent increase in iron-deficiency anemia.

But if farm societies were short on quality, they were long on quantity because of the productivity of their concentrated fields, and this, coupled with better techniques for storing grain and protecting it from rot and rodents, allowed farmers in good harvest years to contribute their surplus grains to a communal grain supply. This in turn fed the development of two characteristics that were carried over from late-on hunting societies—division of labor and hierarchy—but now in just a few millennia came to have an increasingly decisive economic and social role in the life of agriculturalists. Because of surpluses there could be full-time artisans and potters who would be supported from the communal granary, full-time shamans, full-time laborers for dams and irrigation ditches, full-time guards to protect the village from predators human and otherwise, full-time accountants to regulate the collection, storage, and distribution of grain—and full-time rulers, the high-status chieftains who would have had to try to bring order to such a complicated society and see that all these tasks were efficiently done. The hierarchy that had been resorted to in the hunter-gatherer world at times of stress, as for example at Sungir, now evolved into a full-blown, stratified "class society."

Countless sites testify to these developments as agricultural societies matured in the years after 10,000 years ago. This is the long road to what is called civilization, and all the horrors of that are too well known to be covered here: gated cities, armed conflict between settlements, poverty among vast riches, hunger amid plenty, slavery, human sacrifice, aristocracy and royalty, religious despotism, autocratic states, empires, and the despoliation of lands and waters that is inescapably included in every imperial venture. (It is not an accident that all early creation myths, such as the Babylonian *Enuma Elish*, involve original acts of violence and celebrate an ideology of conquest.) As Jared Diamond has neatly summed up the whole process of the origin of agriculture and its aftermath: "Forced to choose between limiting population or trying to increase food production, we chose the latter and ended up with starvation, warfare, and tyranny." It was another fateful chapter in human domination.

It might be wondered, then, why agriculture persisted and indeed expanded across the globe.

The easy answer is that it was a powerful ecological, social, economic, and eventually political force that once started could not be easily contained. It filled a void that nothing else could fill, once the sedentary cultures decided that the supremacy of the human species meant that it was entitled to multiply its numbers as fully as possible and extend its range as far as possible, rather than limit either. And indeed agriculture worked synergistically with that sense of supremacy, for it was a demonstration of human superiority over the species of nature (some of them, at any rate) successful enough to prove that it was the rightful and legitimate pathway of human culture.

And for a time, at least, it fulfilled its promise. It *did* enable humans to survive when food resources were scarce or dwindling, and to expand their range and dominance. Agricultural societies were forced continually to extend their ranges, first to provide new acreage for children setting up their own families, then after a while to find new lands when the original soils were exhausted or the rivers drained dry. If the new territory was unoccupied the farmers would simply settle and begin the process over again; if it was home to hunter-gatherers they would take the land by force, because they had far greater numbers and brought with them diseases deadly to people who had no immunity to them; if it was occupied by another farming society, then the people who depended on the subjugation of animals simply attempted the subjugation of strangers: war.

But in the end agriculture always failed. It was an environmental assault on the earth that was almost never sustainable for much more than a few centuries without disruption and devastation: in the long history of empires dependent on agriculture and irrigation (Akkad, Babylonia, Sumeria, Assyria, Carthage, Mesopotamia, Egypt, Inca, and Aztec among them) we may read the story over and over again: of the exhaustion and salinization of the land, the destruction of forests, the overgrazing of fields, the compaction of soils, the extinction of wild animals, the silting and salting of rivers, the alteration of climate, erosion, desertification—*and*, as agriculture and its attendant systems began to fail, the revolt of the underclasses, or the collapse of the imperial systems, or the invasion of outsiders, or often all three. Nature always ended up having her revenge: of all the places where agriculture started, only one, eastern central China, remains a productive agricultural area today; the rest are deserts or jungles.

That might be a caution, if our species could understand caution. For as the story of agriculture makes clear, the domination of the earth can come only at a price, and as we can tell today the price may well be the despoliation of the earth and the destruction of human systems, perhaps the decimation of the species itself.

FOUR

The Erectus Alternative

1,800,000–30,000 YEARS AGO

What, then, of Eden? Of the time before the Sapiens and their culture of domination, thrust out of the garden and given "every living thing [as] meat for you," alienated from the natural world and confronting it by increasingly powerful weapons and strategies of death?

Before there was *Homo sapiens* there was *Homo erectus*, not a *modern* human exactly but distinctly and unmistakably human nonetheless. *Homo erectus*, so named because it was the first to walk regularly upright, emerged in Africa about 1.8 million years ago, expanded into the Levant and Asia as far as modern China, evolved into *Homo neanderthalensis* in Europe after 250,000 years ago, into *Homo sapiens* in Africa after 200,000 years, and continued on in Asia until about 30,000 years. It lived in all for about one and a half million years, an extraordinary stretch of time, eight times longer than the Sapiens have known, twenty times longer than the Sapiens of modern culture, a period of longevity unrivaled in hominid evolution.[1]

To get some idea of how people lived in this long Erectus era, let us look at one of the most complete Erectus skeletons, the famous Turkana Boy, found at a place called Nariokotome just west of Lake Turkana in northern Kenya, over a course of diggings from 1984 to 1988. When the sixty-seven bones were put together, there emerged the clear figure of a young man who would have been about eleven years old (possibly younger), was 5 feet 3 inches tall, calculated to grow to an adult size of 6 feet 1 inch, remarkably tall for an early human, and would have weighed a little over 100 pounds, growing eventually to 150 pounds in adulthood. He died face down in a marshy bog, 1.53 million years ago—but his bones can speak to us today as if he were only recently dead.

For one thing, such a long, slender stature tells us that he was evolved for a hot, dry, open environment, like the African savanna as it would have been in the warmer climate that the world underwent after about 1.8 million years ago, for he would have had far less of his body open to the direct sun than the stooping apes or his hominid predecessors, and he would have had a greater total of bodily surface exposed so as to dissipate heat, like modern African tribes in this same area, only a few hundred miles from the equator. That would also mean that he and his Erectus forebears had lost the body hair or fur that the more ancient hominids possessed, keeping only some on the top of the head for solar protection, and that he would probably have darkly pigmented skin for the same reason.

For another, the skeleton tells us that it was a fully upright species, well adapted to walking efficiently—and even running, which earlier hominids could not have done. The narrow pelvis would allow the leg muscles greater space and efficiency in movement, and the rib cage was shorter and narrower at the base than in earlier hominids to facilitate movement by these same muscles, compensated for by a wider, barrel-shaped chest to maintain sufficient volume for the lungs. The arm bones too are shorter than in tree-climbing species, indicating regular bipedalism for the first time in the human record, an adaptation undoubtedly necessary for survival on the open savannas. Together this physiognomy indicates that the Erectus bands could have covered an extensive territory—a wide "day-range," as the anthropologists call it—and would have been able to forage for a great many kinds of animals and plants.

But the most interesting thing about the narrow pelvic area is that in the female Erectus—Turkana Boy's mother, let us say—it would have meant a much more restricted birth canal than in earlier hominids, perhaps even a little more constricted than in modern humans, hence a much earlier birth and smaller-sized head at birth so that the opening could accommodate it. That in turn would mean a much longer period of dependency for the child, and parental obligation for the mother, a feature unique to humans, but it would also mean that the child's brain would grow while it was *outside* the womb, open to all the stimuli of the environment and thus greatly increased in intelligence. One study has calculated that Erectus individuals would have tripled the size of their brains after birth, just about as modern humans do, and so have had many of the same capacities as a modern brain though with only a little more than half the size. Erectus brain size averages out to about 1,000 cubic centimeters (Turkana Boy's was calculated at 880 cc., suggesting 906 cc. at adulthood), which is more than 30 per cent greater than the nearest hominid predecessors, and it would grow to 1150 cc. over time. (The early Sapiens had brain sizes on average of 1,230 cc., and modern Sapiens about 1450 cc.)

Such a comparatively large brain would require an enhanced supply of energy from high-quality food (brain tissue uses twenty-two times more energy than muscle tissue), and so too would the lengthy period of nursing during the long Erectus infancy. As Robert Martin of the Institute of Anthropology in Zurich has shown, this would mean the regular consumption of fair amounts of meat, which is a concentrated and nutritionally rich source of calories, protein, and fat, and this is exactly the kind of diet indicated by two parts of the Turkana skeleton. First, because of the narrow pelvis and high chest, the intestinal area is much smaller than in apes and early hominids, suggesting that the Erectus body must have made up in quality of food what it lacked in quantity to process it. That in turn implies a significant amount of dietary meat and marrow, which can be readily broken down in the gut and supplies high value in small quantities, and tubers and vegetables that are processed before being eaten to make them easily digestible. Next, the teeth, particularly the cheek teeth, are much smaller than in the predecessor species and the molars have less protective enamel, arguing that less heavy chewing was necessary, again a sure sign of processed foods—*and*, very likely, cooking.

Evidence for Erectus use of fire is sketchy, and some claim that there is no good evidence before 790,000 years ago (Gesher Beneot Yaakov, in the Levant) or even 500,000 years ago (Zhoukoudian, in China). But a recent elaborate study of a site called, with anthropological arcanity, FxJj20, at Koobi Fora on the other side of Lake Turkana from the Turkana Boy's home, run by a team led by the anthropologist Ralph Rowlett of the University of Missiouri at Columbia, has shown that Africans really did have control of fire at about 1.6 million years ago, and their fires might well have been seen against the nighttime skies by the Turkana Boy family just twenty-five miles across the lake. The evidence is three round, basin-shaped, reddish spots, about twenty inches in diameter, arranged in an arc suggesting a circular encampment, where fires were set not once but many times over (according to modern archeomagnetic tests). To determine whether these were deliberately set by humans and not caused by lightning, the excavators searched for microscopic bits of silica in the round patches produced by living plants and trees that would remain behind when burned; at one patch they found only a single type of silica, indicating that it was probably the residue of a tree that had burned down in a grass fire, but at the three basin-shaped spots they found a "mixed bag" of silicas, "as wood from different trees, as well as grass and other tinder," were fed into the fire. And to confirm human control, 160 flint flakes, some sizable, were found near each of the three fireplaces—and flint scraped against flint will produce sparks that could ignite any tinder.

Fireplaces: now the Erectus bands could deter predatory animals at night, enjoy warmth on the chillier nights—and cook. The Rowlett team has not yet found evidence that tubers and other plant foods were cooked at this site, or at the numerous other basin-shaped spots that have been found in the area, but a stone tool with a serrated edge found in one of the spots shows microscopic "use-wear" evidence of having been used to cut soft tissues, animal or vegetal, and nearby animal bones show cut-marks as if they had been butchered. And it seems only logical that Erectus would have cooked foods, especially varieties of tubers, since they would have been numerous on the savanna in the drier climates after 1.8 million years ago, they are high in energy that the large-bodied Erectus population would have needed regularly, and many of them are edible only when cooked.

The Harvard anthropologist Richard Wrangham and his colleagues have argued that "strong signals" for cooking in the fossil record are indicated by the smaller Erectus teeth and larger Erectus bodies, and they suggest that this was the reason for one of the most important, and unusual, features of human society, beginning with Erectus society: pair-bonding. They reason that with the advent of cooking, food had to be taken to the hearth site instead of being consumed where it was found, and these accumulations could have been forcibly seized or stolen from females by males unwilling to do their own foraging—unless the females established a protective relationship with a male who would fend off competitors. Such a relationship would likely be sealed by females extending their period of sexual availability to the favored co-defender, leading to increased mating with a single partner, and hence a lessening of sexual competition among males. All of which is lent convincing support by the relatively small size difference between Erectus men and women as compared to earlier hominids and most other primates, whose males compete regularly for sexual favors and are a great deal larger than females (for gorillas two times larger). Among other primates where the sex size difference is small or nonexistent (gibbons, some lemurs), monogamous pair-bonding is the rule for long periods, and this suggests that Erectus society, as Richard Klein of Stanford has put it, saw "the beginnings of the distinctively human pattern of sharing and cooperation between the sexes" that is still found in present-day savanna tribes—and, it should be added, still in a few other modern societies as well.

But of course such human pair-bonding is exactly what one would expect, given an early birth and long childhood, which need an effort of great involvement and duration on the part of both parents if their offspring are to survive. Both would have to take responsibility for providing the kind of high-quality food that their children (and, given that larger brain, they themselves) would need, both would have to guard against the ever-present danger of the passing hyena or sabertooth cat. And what it all adds up to is a *family*, an institution particularly human.

The last piece of information that Turkana Boy can tell us is that the Erectus body was very robust and extremely powerful, with thick bones, rugged lower limbs, and powerful muscle markings on the arm bones, indicating, as

Klein puts it, "phenomenal upper body strength." Alan Walker, an anthropologist who was part of the team that originally discovered the skeleton, conducted a survey of thigh bones for all hominid species up to the modern and found that the Erectus bone was far and away the most robust: "Even as an adolescent," he reported, Turkana Boy "maintained remarkably high activity levels [and] as a result, he attained exceptional strength relative to modern humans." And still another study that measured "stature" from the length of thigh bones found that Erectus outscored all other earlier species by 10 to 20 per cent and even beat out modern Sapiens by several per cent.

This in turn indicates that Erectus bands must have led fairly vigorous and active lives, were able to travel great distances, and depended more on bodily strength than on any sort of technology to make their livings on the savanna. They certainly had stone tools, with handaxes, cleavers, knives, flakes, and scrapers that were much more efficient than the tools of their predecessors, and these artifacts were so capable at the tasks to which they were put that they endured for an amazing stretch of 1.4 million years substantially unchanged. But tools did not play nearly as much of a part in Erectus society as they were to play in Sapiens society, nor was there anything like the variety that the Sapiens came up with. Apparently there was no felt need in all that time for crafting anything more complex: the basic handaxes and flakes did the job—probably mostly in food preparation, cutting meat from the bone, smashing bone for marrow, peeling and crushing tubers, starting fires—so why would anyone want anything different? No technology trap here.

And it's not that these tools were unsophisticated. Most of the bifacial handaxes, for example, have a strange beauty to them: they are carved carefully so that they are symmetrical, the flakes removed in an even pattern, so efficiently knapped that it is obvious they were not, as some recent archeologists have claimed, merely rock cores from which flakes were removed. It is no small thing to create such an artifact, for it involves a set of ideas that represent considerable intelligence in this species, beyond the mere feat of planning out a series of acts and learning how to carry them through. Thomas Wynn, an anthropologist at the University of Colorado who has made a careful study of Erectus tool making, argues that it required concepts of design, order, sequence, separation, symmetry, diameter, parallelism, three-dimensionality,

9. Erectus handaxe, Early Acheulean period. Courtesy of John Reader / Science Sources / Photo Researchers Inc.

and the whole basic notion of standardization, all in all a process requiring "a frame of reference of considerable complexity." We are so used to most of these concepts that we hardly appreciate what an extraordinary advance in human thinking they represented. Symmetry, for example, is so frequent in nature, and in the human form itself, that it might seem an obvious concept, and yet, as Wynn points out, it is not an idea that young children seem to have, and it takes a certain developed view of objects to perceive it and then to conceptualize it, a significant intellectual achievement. So too, the idea of perspective—which, mind you, was not worked out in western art until the fourteenth century—and the development of skills to produce it in artifacts, which Wynn argues is a product of "sophisticated notions of perspective and of general frames of reference." All in all, Wynn finds that the manufacturing of such tools, at least by 300,000 years ago, showed that by then "hominids possessed an essentially modern intelligence" long before the Sapiens came along.

One point about these Erectus tools is especially interesting. Although the human brain increased in size over the 1.8 million-year length of Erectus's time on earth, from the 910 cc. braincase of the Turkana Boy's contemporaries to the 1185 cc. of a skull at Ceprano, Italy, dating to about 700,000 years ago—and then to the 1230 cc. of early *Homo sapiens* in Europe—there was no appreciable change in technology in all that time. What caused the increase is open to argument, and much of that has gone on in recent years, but it clearly was nothing having to do with technology, as is sometimes claimed. The most parsimonious explanation, as the anthropologists say, is what Darwin identified as "sexual selection," people choosing mates of greater intelligence over the millennia not merely for more successful procuring of food or protection against predators but for companionship and entertainment around the campfire. Given the enormous size of the human brain, requiring a prodigious amount of energy and the effort to supply it, one seems justified in regarding it as our peacock's tail—at least before ornaments.

So the bones speak: *Homo erectus*, marvelously adapted for life on the African savanna, tall and immensely strong, walking upright, traveling far, with large brains, rich diets, cooking hearths, pair-bonding bands, simple and efficient technology—and nearly two million years of success.

It was undoubtedly that successful lifeway that led to the wide and fairly rapid dispersion of the species out from the savanna to fill much of the African continent—there are nearly fifty sites with Erectus bones or artifacts from modern Morocco to Kenya to South Africa—and then to move outward to the Levant and Georgia in western Asia, India, Java, and China in the east, and eventually to Italy and Spain in Europe, and even Japan. These things are tricky, but the general estimate is that there may have been some ten thousand tribes in all by about 500,000 years ago, though the tribal unit was probably smaller than in the contemporary ethnographic record, and the total population something around 1.5 million.[2] Whether the migrations were push or pull it is hard to tell, but it is generally thought that the chief reason would have been population pressure and a strain on carrying capacity, in many cases made worse by climate changes that forced choice food animals to leave the area or dried up local water holes.

In any case, this dispersal of some populations from Africa after 1.8 million years or so, while many remained behind in the more favorable regions, is a far more wondrous achievement than most paleologists ever acknowledge. To make the decision to leave home, especially if hiving off from a band of family and friends that has grown too large for the immediate resources (or too contentious for social harmony), must be a wrenching course of action, taken as a profoundly generous sacrifice for the sake of others who might never be seen again—altruism without any sure reciprocity. And deciding *where* to go would be an absolute life-and-death task, needing some certain knowledge of distant territory, its climate and food sources and water supply; this would be made easier if the migration were to an area calculated to be roughly similar, as up a major river or along a seacoast, where marine animals would likely be familiar, but still it would be a frightening and no doubt contentious process. Yet it was done, regularly, during the long Erectus reign: this was a human of some complexity, depth, and wisdom, not to mention courage.

Not far from Turkana Boy's home, near the Karari Escarpment about fifteen miles from the lake, is an area known as Site 50. Over three years a team of scientists led by the archeologist Glynn Isaac unearthed 1,405 pieces of stone artifacts and 2,100 fragments of animal bones—antelope, giraffe, hippopotamus, and something zebra-like—that had been deposited there some 1.5 million years ago. Those remains speak also.

Isaac's team discovered that some of the bone fragments bore little V-shaped grooves that experiments showed had been cut by a stone implement, probably a handaxe or maybe a lava flake. This was proof that the bones and stones were not accidentally mingled but that some Erectus group had deliberately brought the bones there with meat on them and butchered them, dismembering carcasses and cutting meat from the shafts. Some of the longer bones had also been shattered, further experiments showed, by being placed on a stone anvil and struck with an implement, again most likely a handaxe, and most probably so that the marrow inside could be eaten. Several of the stone flakes showed signs of having been used for butchering, as well as for whittling wood and cutting plant tissues.

Not much, but it can provide a rough picture of Erectus social life. The most important inference is that these people brought their animal food from the outside to a central place where it was processed and undoubtedly consumed: a home base. (Of course they could do this because they were the first primates to walk upright and had hands free to carry things.) Whether there was a campfire or not, they gathered to butcher and eat in a group, to interact, to laugh, maybe to sing, and, at least rudimentarily, to converse. (Modern consensus is that *Homo erectus* did not have enough neural cells in its relatively small spinal column for the kind of control over the thorax that is necessary for articulate speech, but that does not mean that Erectus had no sounds to communicate with, as all primates do.) We know from the body sizes of the sexes that the males were not competitive, and judging from primates with similar arrangements they were probably cooperative, so it is very likely that food would have been shared among them and their families—since sharing does not fossilize we have to make an assumption, but it is common enough in the ethnographic record. It is even more likely if we realize that the men were probably related, brothers and uncles and cousins, again another feature common among primate groups, where males provide the generational coherence of the band and females are often brought in from other bands.

There is no knowing the exact size of these bands, but anthropologists guided by the size of the encampments and the number of artifacts scattered around—not just at Site 50 but also the seventy or so known Erectus sites elsewhere—generally estimate around twenty-five individuals, five or six families, providing a dozen or so adults for food gathering and prey chasing. This is a recurring number also in ethnographic studies, and it seems to be the size at which a group of humans can cohere best, with enough contacts for a full social life and not too many for social overload; in a group of six people there are fifteen possible one-on-one relationships, in a group of twenty-five there are three hundred; that seems about the limit for daily social interactions in a harmonious group, and would be especially so for Erectus bands with brains smaller than those of moderns.

There is no doubt that the interactions were intense, as they are among all primates except orangutans, but they would have included the special interaction of attending to children who were born too early to fend for themselves and needed the care and attention not only of parents but of relatives and elders, not only in babyhood but for a long period of vulnerability until adolescence. No preceding species had anything like it. Alan Walker, the Turkana Boy anthropologist, has said that this long period of dependency, and the multiple relationships that it would have necessitated, convinced him "that *Homo erectus* was an intensely social creature, with strong cooperative ties to others of its species."

One nice measure of the sociability of the Erectus bands comes from another fairly complete Erectus skeleton found in pieces at Koobi Fora in 1973 and assembled, again with Alan Walker supervising, over the next two years into what became "KNM-ER 1808" at the Kenya National Museum. It is a skeleton 1.7 million years old of an adult female who shows signs of having had a crippling case of vitamin A poisoning, a disease that causes the flesh to separate from the bones and creates blood clots that can then ossify and cause terrible, often immobilizing, limb pains. This woman, despite the agony she must have been in, lasted for weeks or maybe months while her bones thickened all over her body, according to Walker, who suddenly realized: "*Someone else took care of her*. Alone, unable to move, delirious, in pain, 1808 wouldn't have lasted two days in the African bush, much less the length of time her skeleton told us she had lived. Someone else brought her water and probably food; unless 1808 lay terribly close to a water source, that meant her helper had some kind of a receptacle to carry water in. And someone else protected her from hyenas, lions, and jackals on the prowl for a tasty morsel that could not run away. Someone else, I couldn't help thinking, sat with her through the long, dark African nights for no good reason except human concern." This, Walker knew, was "poignant testimony to the beginnings of sociality," the creation of bonds between and among people of a kind not found in any other primate groups, uniquely human and uniquely powerful as a social glue and guide. One reason, surely, for the longevity of the Erectus species.

One other note from 1808. Her vitamin A poisoning, Walker and his team deduced, was caused by an excess of carnivore liver. Carnivores eat lots of livers of other animals, which retain and do not process vitamin A, and their livers become storehouses of it, making them fatal for a human who would eat anything like a pound of them, as 1808 apparently did. Just how she acquired so much carnivore liver is something of a mystery, since the animal that killed the carnivore would likely have eaten all the inner meats first, but perhaps the animals had developed an instinct against eating liver. Who knows when *Erectus* figured all that out, but there are no other known Erectus skeletons with this disease (from a fairly small sample, true, bones of approximately seventy-five individuals) and obviously at some point Erectus societies—the enduring ones, anyway—developed signals or maxims that embodied warnings against liver eating, just as Arctic tribes, for example, have done over the centuries. The tale of 1808 is at any rate proof that these people ate meat, if sometimes the wrong kind—confirmed by her teeth, showing marks of "gouging and battering" of a kind that only meat- and bone-chewing carnivores like hyenas have.

It is maddening that the fossils of this era so long ago tell us so much of who the Erectus were but do not tell us more. To fill out the picture it is necessary to make some extrapolations, as anthropologists usually do, from contemporary and historical tribes presumed to live pretty much as the Erectus did. These are the ones that somehow lasted through the eons, becoming hunters as all Sapiens tribes apparently did, but resisting the tendency toward storage and sedentary life, refusing to succumb to agricultural and pastoral lifeways that surrounded them, keeping even industrialism largely at bay.

The literature on such tribes—living in small bands, with pair-bonding, food sharing, simple tools, and wide "day-ranges" in particular—is copious. (Most often the tribes include the !Kung/San, Mbuti, Baka, and Hadza of Africa, the Andaman Islanders of the Indian Ocean, the Batek of Malaysia, the Pandaram and Paliyan of India, and the Kogi of South America, but there are others.) These tribes have been labeled "immediate-return" societies by the British anthropologist James Woodburn, by which he means that their members, having no way of storing food on the hot savanna, eat what they find on the same day or over a few days, they parcel it out equally, with (in economic terms) distribution having nothing to do with production, and no one goes hungry. This in turn inevitably means that there is no hoarding, and no means of distinguishing one person or family above another, reinforced by strict codes of sharing, a social arrangement with "a remarkable degree of equality" that "systematically eliminate[s] distinctions . . . of wealth, of power and of status." No leaders, no shamans, no chiefs.

Such people also have no dwellings, no permanent camps (though several camps might be visited annually), no real property, and only a vague definition of their home territory, and as a result can move around with great ease, especially useful in times of scarcity, and in fluid groupings that might change a few times a year. Because they are nomadic, and egalitarian, there is no way to accumulate goods or foods, and in any case there is not much in the way of material objects in their lives—even tools are seldom carried from place to place, since they can usually be made from local materials wherever the band travels next. (And among the Erectus there were not even ornaments or clothing, which they did not possess.) Thus there is no sense of jealousy, at least of material things, since there are so few of them and they are generally shared, and none of the disputes common in sedentary surplus-and-storage societies.

Harmony is kept in immediate-return tribes in a number of other ways. Between mates it is fostered by according high stature to women within the band, for not only are they the miraculous producers of the offspring who are the next generation but they generally are responsible for the "gatherer" part of the diet—and plants and nuts make up some two-thirds of the total intake in almost all tribes. Between males harmony is usually ensured by the equality of position and power within a band, reinforced by the cooperation among the men necessitated in the hunt and by the family ties between so many of them; the Mbuti of the Congo, it is said, "look on any form of violence between one person and another with great abhorrence and distaste, and never represent it in their dancing or playacting." And lastly, disputes among families and groups that become threatening are solved with an accepted and indeed honorable tradition of "fission," the hiving-off of individuals or groups that are so disgruntled with their fellows that they can no longer live with them and go in search of another band.

One last feature of these peoples is important to observe: they do not have graves or the religion and magic that go with a belief in spirits and afterlives and the like. (Nor, of course, are there gravesites associated with Erectus remains.) They accept death as part of the normal cycle of life and do not rig rituals that try to deny or extend it to some otherworld: the Hadza leave the bodies of the dead out to rot, while the Baka leave them to be eaten, and say, "When you're dead you're dead and that's the end of you." Similarly, they do not have magical rites that evoke spirits and heavens, for the truly sacred part of life is the natural world in which they live, of which they are an intimate part. For the Mbuti of the Congo, as the anthropologist Colin Turnbull has so effectively shown, the forest around them is the only sacred "presence" they understand, communion with it is all they know by way of spirituality, and magic has nothing to do with it. The forest for them has a life force, a *pepo*, or breath, or wind, that abides in all parts of nature, animate and inanimate, and Turnbull analyzes how this affects their worldview: "The perennial certainty of economic sufficiency, the general lack of crisis in their lives, all lead the Mbuti to the conviction that the forest, regarded as the source of *pepo* and of their whole existence, is benevolent, and that the natural course of life is good. The absence of magic . . . is simply an indication of the normal absence of crisis. And when there is a crisis, they believe that the forest has just fallen asleep, so they sing songs to it to wake it up."

Thus the ethnographic record provides a striking picture of the coherence and concord of immediate-return societies, who are after all managing for the most part (where they have not been severely disrupted by agricultural and industrial invasions) to carry on traditions that in many respects date back more than a million years. Back to the Erectus, who, if we accept the premise that all immediate-return societies must look pretty much alike, can be regarded as having lived in some generally similar way.

The worldview of the Mbutis: that is what I take to be the worldview of Erectus.

It is in its attitude toward nature, as I see it, that the Erectus culture stands farthest apart from the developed Sapiens culture. It would have included, as with the Mbutis, a sense of the overriding benevolence of nature, which after all returned enough to them to make them successful and populous over an immensely long period, and an immediate identity with the space they moved over and the animals they moved among. And I would argue that at the heart of it, the very reason for its benignity, was that Erectus societies did not get the bulk of their meat by confrontational hunting, the bloody and essentially cruel killing of fellow creatures, but rather by scavenging. As a result they did not develop a distance, a mental separation, from the animals, as the Sapiens did—there was no need to think of Self and Other, and to act it out by committing a daily assault on the web of the ecosystem. To kill large mammals.

As I noted earlier, the evidence for hunting of any kind is particularly slim before about 70,000 years ago, and of large animals I would say nonexistent. Hunting, after all, is a complex and difficult way to get the evening meal, demanding considerable practice and planning, and when dealing with most animals with evolved defenses a particularly dangerous way. It is hard to see how an Erectus band would have been able to hunt elephant, hippopotamus, buffalo, and the like—animals whose bones are found at their campsites— especially with the simple handaxes that are the only tools they had that could have been used as a weapon. The idea of a half-dozen men, even as strong as Erectus men would have been, charging at an elephant or a hippo and try- ing to kill it with a hand-held blade and not be trampled is simply ludicrous, as is the notion of men racing across the savanna trying to catch zebras and antelopes and other fleet-footed prey. Richard Klein's sensible conclusion is that "humans had relatively limited ability to hunt large animals" until they developed hafted weapons and projectile points.

And, basically, limited *need*. Klein points out that most Erectus sites in Africa where tools and animal bones have been found—at the Olduvai gorge, Koobi Fora, and Duinefontaine, for example—were not caves but open-air "water-edge" sites by rivers, lakes, and bogs, where all kinds of animals would come to drink, predators made their kills, and "where scavenging opportuni- ties may have been particularly abundant"; in like places in East Africa today, he notes, "substantial amounts of scavengable meat and marrow bone" are left behind and escape the attention of hyenas and their kind, especially in the drier months. And it is perfectly likely that if need be, a group of Erectus would be able to drive most other scavengers off a choice carcass, given their comparatively large stature (even a large hyena is only three feet high at the shoulder) and strength; some tribes, like the Hadza, practice that today.

Besides, since meat probably made up only a third of the Erectus diet (though an important third), as in modern immediate-return tribes, it would not be necessary to scavenge all that much in an ecosystem rich in edible plants and tubers.[3] A band of twenty-five people needing a total of 50,000 calories a day, say, could be expected to get 35,000 of them from plants and would need only 15,000 from meat. A full-grown saiga antelope at 115 pounds contains about 80,000 calories of meat, and assuming that the lioness left just a fifth of the carcass, that would be enough for a band for a day. Scavenging, in other words, would be both a simple and a sufficient solution to the acquisition of meat.

More than that, though the fossil record cannot tell us for certain, there are no grounds to think there would be a *reason* for Erectus to behave any differently from the other animals of the wild savanna, all of which pass each other regularly through the day without flight or fear and drink at the same waterhole at sunset. As scavengers, they could live intimately in nature, as animals and birds do, as hunters necessarily cannot once their killing habits are known. They would know themselves on the animal level, as just another species making its way in the world, part of a seamless membrane of life. There would be no separate *self*, no declaration of individuality as in ocher-decoration and ornamentation, none of which is found in the Erectus record.

That is to say, they would have an Erectus *consciousness* different in some fundamental ways from that of the Sapiens, different from our own. The British philosopher Owen Barfield wrote a book in the 1950s called *Saving the Appearances* that was essentially an attempt to examine what he saw as "the evolution of consciousness," starting with the earliest humans: "a time when man — not only as a body, but also as a soul — was a part of nature in a way which we today, of course, find it difficult to conceive." He regarded these humans as having what he called "original participation," which he said was a "primal unity of mind and nature, with no separation between inner and outer worlds," and no sense of the self apart from the rest of existence — nor, presumably, even a sense of "the rest of existence." As difficult as it is for us to imagine now, Barfield held that there was a time "when men and nature were one in a way that has long since ceased," which we recognize today only in our concept of paganism, "the other name for original participation, in all its long-hidden, in all its diluted forms." That was the reality at the core of Erectus existence.

The science writer James Shreeve comes at this idea from a different but similar perspective. After years of studying Neandertal culture for his book *The Neandertal Enigma*, he believes he has a sense of what the society was like, and it seems very much like what I am suggesting for Erectus society—which would be logical enough, since as I have indicated Neandertals also lacked religion and serious hunting. He writes:

> I imagine Neandertals possessing a different sort of self and a different kind of consciousness. . . . The borders between the Neandertal and the Neandertal world are fuzzy. For us, consciousness seems like an inner "I" resting somewhere deep in the mind, eavesdropping on our stream of thoughts and perceptions. . . . I would give the Neandertals a fictive inner voice, too, but move it out, away from the center, so that it speaks from nearer to that fuzzy border with the world. A Neandertal thought would be much harder to abstract from the thing or circumstance that the thought is *about*. The perception of a tree in a Neandertal mind feels like the tree; grief over a lost companion *is* the absence and the loss. Neandertal psyche floats on the moment, where the metaphor of consciousness as a moving stream is perfect, the motion serene and unimpeded by countercurrents of re-think, counter-think, and double-think. . . .
>
> The Neandertals' spirit *was* the animal or the grass blade, the thing and its soul perceived as a single vital force, with no need to distinguish them with separate names. Similarly, the absence of artistic expression does not preclude the apprehension of what is artful about the world. Neandertals did not paint their caves with the images of animals. But perhaps they had no need to distill life into representations, because its essences were already revealed to their senses.

Similarly, the San of the Kalahari, according to Laurens van der Post, live in a state of "dependence and interdependence" with nature such that they put themselves into all its elements and "know what it actually felt like to be an elephant, a lion, an antelope, a steenbuck, a lizard, a striped mouse, mantis, baobab tree, yellow-crested cobra or starry-eyed amaryllis."

That worldview, that essential identity with the natural world, that inherent harmony with living creatures, was what made Erectus society so enduring. They were a people who, no matter where in the world, whether in tropical or temporal ecosystems, whether at savanna waterholes, Javan shorelines, or Chinese caves, preserved the earth because they knew they were part of it. For nearly two million years. Domination was not a concept that the Erectus mind could have absorbed, nor would it know how to develop the tools to effect it if it could.

Paul Shepard, the important environmental philosopher, has lauded this "interpenetration with the non-human world [as] an extraordinary achievement of tools, intellectual sophistication, philosophy, and tradition." But in summing up the world of the "pre-agricultural" peoples, he goes even farther:

> It is not only, or even mainly, a matter of how nature is perceived, but of the whole of personal existence, from birth through death. . . . In the bosom of family and society, the life cycle is punctuated by formal, social recognition with its metaphors in the terrain and the plant and animal life. Group size is ideal for human relationships, including vernacular roles for men and women without sexual exploitation. The esteem gained in sharing and giving outweighs the advantages of hoarding. Health is good in terms of diet as well as social relationships.
>
> Organized war and the hounding of nature do not exist. Ecological affinities are stable and non-polluting. Humankind is in the humble position of being small in number, sensitive to the seasons, comfortable as one species in many, with an admirable humility toward the universe.

And that, so I am arguing, was Eden.

Comparatively speaking, of course, and with no attempt to suggest that it was paradise of some Biblical kind. But it *was* a life in many respects in which people gathered their food from a fruitful garden, with "every herb bearing seed," and "every tree in which is a tree yielding seed," and "every green herb for meat," and were content to "dress and keep it" for eons. It was a life of intimate families, pair-bonded mates, sharing and cooperation during the day of foraging and around the campfire at night; of "intensely social" interactions and mutual caring; of sexual and economic equality, without signs of hierarchy or authority; of harmony within the band and with the world around, founded on a unity with nature; and an overarching pervasive understanding that "the natural course of life is good."

Of course there were hardships, and many times of lean and fallow and scarcity, and deaths in childbirth and in tigers' jaws, and spasms of severe weather and periods of adverse climate, and there would have been some who violated the social norms and flouted age-old customs. But Erectus societies could not have been successful for so long in so many parts of the world unless they were doing something right.

In what sense could there be any lessons in all of this for us today?

No, I don't mean that there is any way for us to live an Erectus life in the world as it is now, even if substantial numbers of people wanted to and there were not six billion people covering the earth. The defenders of the present who say "you can't go back" are of course absolutely right, and those who attempted to do so would end up looking like the French aristocrats of the eighteenth century who dressed up as shepherds and milkmaids to celebrate a romantic sense of Nature and the goodness of "primitive" life.

However, in the deepest sense we already *are* back. The Erectus way of life is in some sense encoded in our genes after 1.8 million years, and we have more or less the same genome that the first Sapiens possessed 200,000 years ago; and just as we have the same grasping hands and color vision, so too we have the same brains and psyches, and they can be used to perceive the world the same way. As Carl Jung once said, "Every human being has a two-million-year-old man within himself, and if he loses contact with that two-million-year-old self, he loses his real roots."

That does not mean that we are now about to create a foraging life of small, mobile bands living in harmony with nature all over the world. But it does mean that there are aspects of Erectus (and immediate-return Sapiens) *consciousness* that we are capable of comprehending today and adapting to our present lifeways to one degree or another should we make the effort and put our minds to it, individually and to a certain extent collectively. Shepard has argued that all societies "are composed of elements that are eminently dissectible, portable through time and space," and that it is possible to "go out or back to a culture even if its peoples have vanished, to retrieve a mosaic component, just as you can graft healthy skin to a burned spot." We may not be able to create a full-fledged Erectus society or imitate the Mbuti and the

!Kung, but, he says, "removable elements in those cultures can be recovered or recreated, which fit the predilection of the human genome everywhere."

Thus it is a matter of will, not of ability, and as to the imperative of modern society coming to exert its will in some such direction there is no doubt: "Without a global revolution in the sphere of human consciousness," the Czech statesman Václav Havel told the U.S. Congress in 1990, "nothing will change for the better in the sphere of our Being as humans, and the catastrophe toward which this world is headed, whether it be ecological, social, demographic or a general breakdown of civilization, will be unavoidable." Our Being as such humans as Erectus consciousness might lead us to.

Let me give an example, to suggest that this is not all frivolous. Our contemporary culture, certainly in the West, is saturated with an antagonism to and fear of death, and a fantasy of an afterlife (or botox) to escape it, and much of our time and ingenuity (not to mention money) is devoted to staving it off at any price: medicines, machines, chosen foods and supplements, surgery, hospitals, laboratories, institutions, and agencies. Our religions render death up as a terrible and tragic fate—the Christian one even has a gruesomely murdered man as its central symbol—which can be ameliorated only by a conviction that if you give your life to some God you will overcome death with immortality; our popular culture disgorges images of it incessantly and graphically, often to the point that it comes to seem meaningless and trivial so that we may deny its reality; our media are preoccupied with those who effect and suffer it, by war, accident, homicide, or catastrophe, until it embeds the impression that it is the one terrifying and recurrent reality of daily life. This is not a healthy way to view the world, and it leads to deep pathologies in thought and behavior.

An Erectus worldview would accept death, in the way the immediate-return peoples do, as an ordinary part of life and not endowed with any special fearsome or unusual severity: "When you're dead you're dead." There would be no complicated measures to keep people alive when terminally ill or injured; no second thoughts about putting to sleep an extraordinarily deformed or deficient baby; no laboratories devoted to creating drugs or technologies to extend a normal life; no institutions to keep people alive when they have lost mental or bodily functions: a return, in short, to natural selection. There would also be a great many fewer people, of course, and fewer making an enormous drawdown of money and resources, and hence a more reasonable balance of the human impact on the earth's finite treasures.

"Fewer people" contains other aspects of an Erectus consciousness. It accepts the idea, as the science of ecology phrases it today, of a "carrying capacity," a limit on the human use and depletion of natural resources consistent with the survival and flourishing of all the other species, plant and animal and natural, in a given ecosystem. This will at some point mean the regulation of human population growth, fitting the number of people to the needs (and tasks) of the carrying capacity of the region, so that too many animals are not killed, too many plants uprooted, too many toxins spread, too many systems abused. This was the practice of all successful Erectus societies—if they exceeded the limits they would inevitably be *not* successful, and would have to migrate if they did not perish—and of most simple tribal societies we know about, which is why on the one hand they would prolong breastfeeding, have customs to spread out the years between births, use abortifacient plants when necessary, and let infirm people detach themselves from the village to die, and on the other hand take care to assure that vital species were not overhunted or overharvested.

The ideas of population control and limited consumption—not, by the way, so very radical or difficult to imagine today, despite advanced capital's resistance to them—also link to ideas of *scale*. The scale of Erectus society, as we have seen, centered around bands of about twenty-five people, interacting with perhaps six or seven other bands as an occasional tribal assemblage, and that set into a total known universe of maybe four to five hundred people with the same dialect and culture, the usual size of most historical tribes we know of. That wider number would be enough to provide a sufficient number of mating partners for youngsters reaching maturity—a mating network much smaller than that is likely to produce inbreeding and genetic defects—and to exchange information on environmental or resource crises and means of overcoming them ("This root is good if you scrape off the skin," "There are hartebeests migrating just over the big river"), even to provide a safety net for one band suffering from a local calamity. The range of such a tribal unit might be something on the order of seven thousand square miles, assuming an effective population density of two people per square mile (as among modern Hadzas in Tanzania), about the size of modern Swaziland.

The number five hundred actually seems to be pretty much the maximum level at which people can effectively maintain any kind of intimate network, and our brains may have evolved for two million years to adapt at that level. According to Hans Blumenfeld, who has studied this topic as an urban planner, the level at which "every person knows every other person by face, by voice, and by name" fades out "with much more than a 500 or 600 population." It is a rule of thumb among architects that primary schools be built for no more than five hundred pupils, because that is the limit at which a principal will be able to know every one by name; the scholar John Pfeiffer notes that "the memory capacity of the human brain probably plays a fundamental role of some sort since that influences the number of persons one can know on sight." Tribal societies, Erectus included, clustered around that number for a *reason*. Now I would be the first to acknowledge that resettling modern populations in villages of five hundred people, though consistent with our evolutionary history and mental capacities, might prove difficult to do—only one more proof that the human species has vastly overexpanded itself. But that certainly could be a guide for architects and urban planners, even in setting out units within large cities, and a model for suburban developments and "new cities" projects. And within a society drastically cutting population, as our Erectus heritage suggests is necessary—and our environmental destruction suggests is inevitable—such a resettlement to the scales we are hardwired for does not seem untoward.

Erectus consciousness, then, is approachable in modern terms and possible—and clearly *desirable*—in a good many areas. There are others. The Erectus diet, for example, would serve as one even more appropriate for the modern age than the guidelines suggested by the Agriculture Department, as some 44,000 sites on the internet attest: based on lean (wild) meat, high fiber, a wide range of vegetables and fruit, and no milk, cereals, and sugar, it would be infinitely healthier than the average human intake anywhere in the world today. Or the idea of simple tools and minimal possessions, freeing people from the materialistic trap of modern technology and turning them to what certain of them have been calling (and apparently what millions of them are living) "voluntary simplicity." Or the practice of close mother-child bonding, with lengthy breastfeeding and regular physical intimacy, for long periods of childhood, along with child-rearing shared with grandmothers, aunts, and friends, which many recent studies have shown are not only important but indeed essential—because, after all, they are Erectus-rooted—to both physically and mentally healthy adults; the findings that day-care centers are actually harmful for very young children are consistent with this.

Erectus consciousness in the end is not so foreign, and much of it is nothing more than a guideline to human health and sanity, so deeply rooted in our genes that even now we know it to be true.

But it is in one area in particular that this consciousness is vital above all others for us to attempt to comprehend and absorb within us: our attitude toward nature.

I need not belabor evidence for the attitude toward nature that ancient civilizations everywhere took from the Sapiens hunting culture and the agricultural domesticating culture and forged into philosophies of increasing domination and destruction; nor is it necessary to retell the experience of modern scientific civilization in ratcheting up these philosophies into an underlying ethos and equipping it with habits of thought and tools of power that have enabled us to extend our dominance over the globe with the force of a hundred unhalting hurricanes, determining life and death for all other species and systems we encounter. But it is well to understand the simple truth that western civilization has always defined itself in opposition to nature—as Freud said, civilization was necessary "to defend us against nature," including our own—and to realize that this opposition is critical to the way the civilization views itself and operates. What a terrible indictment, to have a culture that prides itself on its distance from the natural world and the natural cycles and rhythms, that regards its mission as needing (in Francis Bacon's words) "to conquer and subdue" nature with its indomitable technology, and that is built on the idea that nature has value only if it is harnessed and exploited for economic purpose: "Nature, Mr. Allnutt," says the spinster on the *African Queen*, "is what we have been put in this world precisely to rise above."

It might even be said that by the twenty-first century western civilization's opposition to all that is not civilized and domesticated has been so successful that in one sense, as the Duke University literature professor Fredric Jameson has said, "nature is gone for good." By which he means that the instruments of advanced capitalism, including industrialism, commercialism, corporatism, financial markets, agribusiness, tourism, trade, media, advertising, all on an encompassing global scale, have caused, at least on the effective conceptual level, "a radical eclipse of Nature itself." Not only has nature been conquered and plundered and exploited for human ends, it has even been subsumed and cast aside by the dominant culture as an irrelevant image, except in advertisements.

But the fact is that nature is *not* gone from our souls, no matter how much capitalist civilization has distorted and dismantled it or driven it from our daily sensibilities. Not only is there that two-million-year-old primal self within all of us, but it is genetically encoded to understand and appreciate nature the way the Erectus self did, even though it sometimes seems buried under and stifled by the modern self. We evolved in a wilderness of extraordinary diversity where we lived in daily intimacy with animal life and plant variety, upon which we depended completely and unbrokenly for survival, and that has only been reinforced by natural selection through 72,000 generations over the long millennia. Hence under our modern veneer, and in spite of the multiple obfuscations of capitalist culture, we still have an innate need for connections to nature, we have an ineradicable appreciation of its flora and fauna, and we have the capacity and somewhere the felt ability to achieve a communion with beings other than ourselves and settings other than those we create. Edward Wilson, the Harvard biologist, has named this "biophilia," and he says that it is "the innately emotional affiliation of human beings to other living organisms . . . [that is] hereditary and hence part of ultimate human nature."

It is upon this that a modern Erectus consciousness of nature can be built.

It would begin with a basic understanding that nature is good, just as the Mbuti know. That seems simple enough a concept, but it is not one that our culture has fostered: we are taught to know that we have been *expelled* from the good world of the garden into the bad and fierce and wild one of nature—"cursed is the ground for thy sake, in sorrow shalt thou eat of it all the days of thy life"—and we are told that it is our task to subdue it. It is true that in some societies at some times there have been some few who express appreciation for nature and revel in opportunities to see the mountains, hike the trails, raft the rivers, absorb the birdsongs, and gaze on the gazelles, but that is not the same thing as having a deep understanding of nature as a benevolent force and the earth as a living, giving source of all life. Nor does it embody the Mbuti sense of daily life as a rich, full, easy, bountiful endowment of nature, rather than something that must be wrested from the world by human effort in a never-ending battle of drudgery and challenge and competition. Einstein thought it was a profound question whether the universe was friendly or unfriendly—a grounded Erectus-guided soul would not wonder.

Then it might seek a thoroughgoing reintegration with nature, a conscious identity with it and its species, something very like the feeling of loss of self and ego, an at-one-with-the-world sensation, that comes when the right temporal lobe of the brain takes control during deep meditation, or when in Zen Buddhism the initiate is lost in the immediate moment and perceives, nonverbally, the law of interdependence. The psychologist Hans Loewald, like Carl Jung himself, has suggested that the "quest for boundary loss, for the merger of Self and Other," is a fundamental human search, and it may be so because that is the "interpenetration with the non-human world" that the Erectus psyche seems to have known. This has certainly been the goal of some people in our civilization, defying the norm, people like the Welsh writer John Cowper Powys, who once asserted that "we can restore, by means of our imaginative reason, that secret harmony with nature which beasts, birds and plants possess but which our civilization has done so much to eradicate from human feeling. . . . It flows through us, stirred by unexpected little things, a magical rapport, bringing indescribable happiness between the solitary ego and all that we behold on this green earth."

And indeed we all know people who say they have had some such experience, proving that it is inborn in us, only waiting to be released. Ellen LaConte has described one, as she lay in a tree in a deep woods on an August day:

> I yielded completely to desire, and clasped the whole rising and falling Life of the Forest passionately to me as it was given to me in the body of That One Tree.

Soon a delicious tremor claimed me, carrying me to a place so old and deep, a place of such pure erotic sensation, I could not remember its like. . . . The Forest was no longer only "out there," but also within me, in my ears and nostrils, under my skin, in my imagination, on my tongue. . . . I knew the Forest then, or almost *was* it. . . .

I was not so much aware of as *part of* the tree beneath me. My spine repeated its spine, my skin grafted to its skin . . . and then, ever so briefly, I wasn't any more—at least not as "distinct from." I disappeared into tree and Forest as surely as the novice disappears in prayer into the Creator.

This is the elemental, age-old self bursting into the modern consciousness with the power and passion that Erectus life must have been filled with daily.

One more aspect of the Erectus understanding of nature available to us is the essential wisdom of *biocentrism*, a way of coming to regard the human, as the ecotheologian Thomas Berry has put it, "at the species level," as one more creature on the earth in essence no grander and greater than the rest, and at heart ultimately dependent upon them and their continuing healthy interactions for our very lives. We are so cocooned in our human-centeredness in most of our existence that this sort of humility seems well-nigh degrading, or juvenile, but it is of course the crucial element of a worldview that knows domination to be wrong and integration to be right. As Berry has phrased it: "Our secular, rational, industrial society, with its amazing scientific insight and technological skills, has established the first radically anthropocentric society and has thereby broken the primary law of the universe, the law of the integrity of the universe, the law that every component member of the universe should be integral with every other member of the universe and that the primary norm of reality and of value is the universe community itself in its various forms of expression, especially as realized on the planet Earth."

Or as the philosophers of a movement called Deep Ecology have recently put it: "The well-being and flourishing of human and nonhuman life on Earth have value in themselves . . . independent of the usefulness of the nonhuman world for human purposes. Richness and diversity of life forms contribute to the realization of these values and are also values in themselves. Humans have no right to reduce this richness and diversity except to satisfy vital needs." All life is sacred, say the Indians, including the stones and waters and clouds and the earth itself, and there is no hierarchy determining that humans are supreme and can dominate and direct the others. We have lived as if there were one, and now we must live another way. The Erectus way.

It is not so esoteric or arcane after all, the Erectus way. It is clear that a great many people now have, as countless ones did in the past, a pretty clear understanding of what kind of attitude to the world would most likely befit the human species and ensure the harmonious continuance of the rest of life. We need not find the models a million years in the past, for their ideas are available to us in one degree or another today.

The immediate-return tribes, for example, those who have resisted western civilization's fateful embrace and allowed the story of their ancient truths to be copied down. And other aboriginal tribes, including many in North America, though compromised and corrupted by European conquest, that still have remnants who know the old ways and remember the ancestral teachings and honor them as well as they can. They have come a long way from their hunter-gatherer past, most of them, but they retain the fundamental lessons that they learned when they came through the wrenching experience of the period of mammalian extinctions: that humans must step carefully on this earth, with respect for the lives of other species, careful to keep intact ecological integrity and what Aldo Leopold once called "the web of inter-dependencies so intricate as to amaze."

The six-nation Irokwa tribes of eastern North America, who know themselves as the Hau de no sau nee, spoke to the world more than twenty-five years ago, setting out the basic tenets of their culture. It is almost as if the Erectus bands were speaking:

> In the beginning, we were told that the human beings who walk about the Earth have been provided with all the things necessary for life. We were in-structed to carry a love for one another, and to show a great respect for all the beings of this Earth. We are shown that our life exists with the tree life, that our well-being depends on the well-being of the Vegetable Life, that we are close relatives of the four-legged beings. In our ways, spiritual consciousness is the highest form of politics.

Ours is a Way of Life. We believe that all living things are spiritual beings. Spirits can be expressed as energy forms manifested in matter. A blade of grass is an energy form manifested in matter—grass matter. The spirit of the grass is that unseen force which produces the species of grass, and it is manifest to us in the form of real grass. . . .

We believe that all things in the world were created by what the English language forces us to call "Spiritual Beings," including one that we call the Great Creator. All things in this world belong to the Creator and the sprits of the world. We also believe that we are required to honor these beings, in respect to the gift of Life.

Our traditions were such that we were careful not to allow our populations to rise in numbers that would overtax the other forms of life. We practiced strict forms of conservation. . . .

The Hau de no sau nee have no concept of private property. This concept would be a contradiction to a people who believe that the Earth belongs to the Creator. . . .

Before the colonists came, we had no consciousness about a concept of commodities. Everything, even the things we make, belong to the Creators of Life and are to be returned ceremonially, and in reality, to the owners. Our people live a simple life, one unencumbered by the need of endless material commodities. The fact that their needs are few means that all the peoples' needs are easily met. It is also true that our means of distribution is an eminently fair process, one in which all of the people share in all the material wealth all of the time.

Ours was a wealthy society. No one suffered from want. All had the right to food, clothing, and shelter. All shared in the bounty of the spiritual ceremonies and the Natural World. . . . All in all, before the colonists came, ours was a beautiful and rewarding Way of Life.

Or, on a different level, there are many strains in the worldwide environmental movement as it has developed over the last fifty years that have tried to enunciate, and in many cases achieve, a consciousness of nature quite at odds with the western paradigm. (Not surprising, because what James Parks Morton of the Interfaith Center of New York has called the "dawning ecological consciousness" is, as he says, "the crown and climax" of the world's spirituality and has created "the revolution in religion itself.") There is, for example, a perspective that has been called "the New Cosmology," a way of understanding the universe as an unfolding story of human and cosmic alignment. Brian Swimme and Thomas Berry have tried to fashion the tenets of this idea, but in the end it comes down to a spiritual appreciation of the totality of life on earth and the realization that the "universe community" is made up of equally worthy beings. The concept crafted by the British scientist James Lovelock in his Gaia Hypothesis, that the planet Earth is literally alive, purposeful in some sense and self-controlling and self-regulating, fits into this cosmology neatly, and that too is a widely adopted idea within the environmental movement and beyond. It is really the old idea of the *anima mundi*, the soul of the earth, reified through science.

Other strains include the Deep Ecology followers already mentioned, who in a great number of books and articles (more than a thousand genuine cites on Google) have articulated a profound ecological ideology that seeks, as the former City College of New York philosopher Andrew McLaughlin sums it up, "a transformation of people and society, advocating the joys of an expanded sense of identification with nature." Of a similar nature is the bioregional movement, centered in North America and with more than 120 affiliate groups, which advocates a decentralization of political and economic life to the level of regions defined by nature (flora, fauna, hydrology, geology, etc.) rather than by legislature, as, most frequently, watersheds and mountains. Explicitly biocentric, the movement has announced itself this way: "Bioregionalism recognizes, nurtures, sustains and celebrates our local connections with: land; plants and animals; rivers, lakes and oceans; air; families, friends and neighbors; community; native traditions; and traditional systems of production and trade."

Certainly some part of the modern animal-rights movement is also biocentric in its assertion of moral rights for all animals, though not much of the debate has promoted the notion of humans being equal in status or value. In fact at least a strong part of the movement operates from ideas of *stewardship*, arguing that since humans are the dominant species on earth it is our moral obligation as stewards, or managers, of it all to treat animals with kindness and decency; as if it weren't our domination that was the problem in the first place. Still, in its straightforward denunciation of "speciesism," the movement begins to approach the consciousness that understands "all living beings are spiritual."

One other interesting force, out of the environmental movement as it is blended with modern-day anarchism, represents a direct philosophical challenge to the basic tenets of western civilization, indeed of civilizations of any stripe. It generally goes by the name of primitivism, though not all anti-civilization types would necessarily define that idea the same way, and for more than thirty years it has put out a string of articles and books (and set up a web site) that on the one hand forthrightly attack civilization in all its guises, and on the other just as forthrightly promote a "primitivist alternative" modeled on prehistoric hunter-gatherers.

The dissent from modern civilization is of course long-standing, and with it have stood such people over the centuries as Schiller, Thoreau, Spengler, Morris, and Mumford, but the present dissenters have assailed a whole array of civilization's underpinnings, from agriculture to industrialism, right to the level of some of its most fundamental modes of thought. John Zerzan, for example, an independent intellectual who has been primitivism's most ardent and articulate spokesman, has made scholarly critiques not only of domestication, religion, psychiatry, postmodernism, progress, and division of labor but, at a deeper level, such cultural fundamentals of civilization as symbolic thought, language, numbering, art, and notions of time. "Why," he asked at one point, "would one respond positively to art? As compensation and palliative, because our relationship to nature and life is so deficient and disallows an authentic one. As [Henry de] Montherlant put it, 'One gives to one's art what one has not been capable of giving to one's own existence.' It is true for artist and audience alike; art, like religion, arises from unsatisfied desire."

And: "Number is the most momentous idea in the history of human thought. Numbering or counting . . . gradually consolidated plurality into quantification, and thereby produced the homogeneous and abstract character of number, which made mathematics possible. From its inception in elementary forms of counting . . . to the Greek idealization of number, an increasingly abstract type of thinking developed. . . . As William James put it, 'the intellectual life of man consists almost wholly in his substitution of a conceptual order for the perceptual order in which his experience originally comes.'"

More generally Zerzan has charged: "Genetic engineering and imminent human cloning are just the most current manifestations of a dynamic of control and domination of nature that humans set in motion 10,000 years ago, when our ancestors began to domesticate animals and plants. In the 400 generations of human existence since then, all of natural life has been penetrated and colonized at the deepest levels, paralleling the controls that have been ever more thoroughly engineered at the social level. Now this trajectory can be seen for what it really is: a transformation that inevitably brought all-enveloping destruction that was in no way necessary."

Interestingly, the prize-winning novelist Daniel Quinn has come at much this same critique with his highly entertaining character Ishmael, a talking gorilla, whose analysis of "the Takers"—civilization, in effect—runs this way: "As I make it out, here are four things the Takers do that are never done in the rest of the community, and these are all fundamental to their civilized system. First, they exterminate their competitors, which is something that never happens in the wild. . . . Next, the Takers systematically destroy their competitors' food to make room for their own. Nothing like this occurs in the natural community. . . . Next, the Takers deny their competitors access to food. . . . [Their] policy is: Every square foot of this planet belongs to us, so if we put it all under cultivation, then all our competitors are just plain out of luck and will have to become extinct." And then his human disciple finally sees the light: "I no longer think of what we're doing as a blunder. We're not destroying the world because we're clumsy. We're destroying the world because we are, in a very literal and deliberate way, at war with it." No surprise that Quinn's first book after his Ishmael novels was entitled *Beyond Civilization.*

The "primitivist alternative" to the deadly mess of civilization is essentially based on the anthropological and paleoarcheological work that began to emerge in the 1960s, highlighted by the "Man the Hunter" conference at the University of Chicago in 1966, transforming the picture of prehistoric cultures from ones of hardship, deprivation, toil, and sickness, "nasty, brutish, and short," into ones of ease, comfort, sufficiency, well-being, and peace, "the state of nature."[4] Since then the bulk of the scholarly literature has reinforced and sharpened that picture, more or less summed up in the *Cambridge Encyclopedia of Hunters and Gatherers* in 1999, and it is this work that informs most of the ideas of what a desirable primitivist society would look like. Marshall Sahlins, for example, an anthropologist whose *Stone Age Economics* has been highly influential in the field and out, has described what he has called "the original affluent society": "Hunting and gathering has all the strengths of its weaknesses. Periodic movement and restraint in wealth and population are at once imperatives of the economic practice and creative adaptations, the kinds of necessities of which virtues are made. Precisely in such a framework, affluence becomes possible. Mobility and moderation put hunters' ends within range of their technical means. An undeveloped mode of production is thus rendered highly effective."

And they do it on "a mean of three to five hours per adult worker per day in food production." And as to poverty: "The world's most primitive people have few possessions, *but they are not poor*. Poverty is not a certain small amount of goods, nor is it just a relation between means and ends; above all it is a relation between people. Poverty is a social status. As such it is the invention of civilization."

It is from this kind of scholarship, which has multiplied in recent years, that primitivists have taken support for the possibility of current alternative living. "Mounting evidence," says Zerzan, "indicates that before the Neolithic shift from a foraging or gatherer-hunter mode of existence to an agricultural lifeway, most people had ample free time, considerable gender autonomy or equality, an ethos of egalitarianism and sharing, and no organized violence. Archeologists continue to uncover examples of how Paleolithic people led mainly peaceful, egalitarian, and healthy lives for about two million years. . . . Ever-growing documentation of human prehistory as a very long period of largely non-alienated life stands in sharp contrast to the increasingly stark failures of untenable modernity."

Some sections of the primitivist movement have even attempted to put some of these lifeways into effect. There are at least a hundred organizations that offer instruction in primitive living—the Teaching Drum Outdoor School in the North Woods of Wisconsin, for example, offers a year-long course in skills for survival in the wild—and in such ways of the American Indians as the vision quest, by which youngsters were initiated into adult society by going alone on a fast into the wilderness and becoming so thoroughly immersed with nature that they would take their subsequent names from an animal or object that they bonded with. A Society of Primitive Technology, begun at the Schiele Museum of Natural History in North Carolina in 1989, now has affiliate schools in thirty states (and Canada and Mexico) that teach "wilderness living and survival," and a biennial bulletin that has reports on making and using all kinds of Stone Age tools, including atlatls, bows, and boomerangs. Attachment to such archaic technology has not prevented the society from launching a web site ("Primitive.org"), nor a similar group from establishing a Primitive Technology Home Page that lists sixty-eight linked groups.

I know of no group of primitivists who have attempted to establish an ongoing Stone Age community of an immediate-return or even full-scale hunter-gatherer type, but then if such a one existed it would likely not wish to have any contact with (or web site for) the rest of the corrupting world. There are a great many individuals, though, who live along the gradations to a primitive lifestyle, using Stone Age tools (and often making them for sale), hunting for food and clothing, living in earth lodges or teepees, making friction fires without matches, weaving baskets and nets, collecting edible wild plants, growing simple vegetables, or foraging roadkill. Some may be trying to prove a point in a political sense, but it seems that most are just living the way their bodies and souls tell them is the healthiest and most satisfying. Says one man who lives in the woods on the banks of the Hudson River, "I'm doing it because I want to live a natural existence outside, and because North American Indians perfected the system of living outdoors on the North American continent. On this landmass at this latitude, I chose the practices best suited for living."

Sounds very Erectus to me.

There is no going back to Eden. Once we have learned of evil as well as good, for which we were expelled, and learned to create the tools that enabled us to carry it out, we cannot entirely abandon them or the habits of thought that accompany them.

But in evolutionary terms, what we have developed now as modern civilization has lasted a mere blip in time, and it is has had very little real depth of influence on our basic hominid nature. Underneath the veneer is a Stone Age mind and a Stone Age heart, and they may still be our guide today, leading us toward a saner and more harmonic world in which the human is in balance with the rest of nature.

For we cannot, we *will* not, go on as we are. In the first place, every civilization that has ever existed has collapsed, and ours will be no exception, because civilizations carry out an iron imperative of ecological destruction, especially intensified by agriculture, combined with social incoherence, especially intensified by hierarchy. Moreover, our particular industrial civilization has developed technologies that enable us to hasten that destruction and incoherence to a degree unimaginable to any preceding one, with consequences so catastrophic that the future of a great many surface species, including the dominant bipedal one, is uncertain. And even if in the next decade or two we do not succeed in altering the climate and atmosphere and befouling the soil and water and air in such a way as to imperil life on earth, there is every reason to expect (as a Pentagon report of February 2004 predicts) a conjunction of crises that will create havoc, war, starvation, disease, and death on a wide scale in every land on earth, and bring our civilization crashing down around our heads.

It is then that we will need the wisdom of the Erectus and the skill of the Sapiens, our Stone Age hearts and minds, to survive.

NOTES

Chapter 1: The Dawn of Modern Culture

1. The taxon *Homo erectus* is a tricky one. Some paleontologists choose to divide it into an Erectus species that goes to Asia and a *Homo ergaster* species that stays in Africa and then becomes *Homo heidelbergensis* at about 500,000 years ago. I find that needlessly complicated and unnecessary. Some variations can indeed be found within these groups, but they are not severe and they do not require a new species name unless some archeologist wants to get credit for finding a new species, as they often do. I favor the older traditional division of *Homo habilis* from 2.5 million years ago to 1.8 million, *Homo erectus* to 500,000 (though continuing in Asia to 30,000), evolving then into *Homo neandertalensis* at about 250,000 and into *Homo sapiens* at about 195,000. Richard Klein, who in his classic textbook uses the taxons *Homo ergaster* and *Homo heidelbergensis* for early and late Erectus, acknowledges that Ergaster "resembled H. erectus in many key features" (Klein, *The Human Career*, 287) and that the derived features the two species shared are "striking" (293), and he ultimately acknowledges that though there are slight cranial differences, the body and limb fossils "fail to differentiate H. ergaster, H. erectus, and the early representatives of H. sapiens" (283).

2. The psychologist Chellis Glendenning has suggested to me that the response of Sapiens to the volcanic winter catastrophe seems very much like that of contemporary people, after a war or natural disaster, who have what the American Psychiatric Association calls "post-traumatic stress disorder." Among its symptoms are "numbing one's responsiveness to the world," which in the Sapiens' case might make them better able to inflict killing on large animals, and at the same time "hyperalertness" to certain stimuli, which might be a useful attribute for the hunt. In addition, she notes, the "psychic distancing" that such traumas create, when "the mental draws up and away from the body," might have led to the symbolic thinking that apparently emerges about the same time.

3. Among the caves with evidence of hunting and fishing in this period are Howieson's Poort, Border, Boomplaas, and Diepkloof caves in an age range after 75,000 years ago, and Klasies River Mouth, Hearths, Nelson Bay, Peers, Apollo 11, Still Bay, Elandsfontein, Sea Harvest, Herold's Bay, Hoedjiespunt, and Boegoeburg around 71,000. The

spear industries named after the Howieson's Poort and Still Bay caves are pretty much alike, though the former include points blunted on one side ("backed" pieces), perhaps to aid in hafting.

4. Some scholars like Alexander Marshack (in many articles and books, including *The Root of Civilisation* (New York: McGraw Hill, 1991) and *Current Anthropology* 37, no. 2:327–37 (1996)) and Robert Bednarik (for example, *Cambridge Archaeological Journal* 21 (1992): 27–57) in particular are able to find earlier indications of human-made signs and artwork almost anywhere. But I would say that generally scholarly opinion sides with Richard Klein's finding that "on many of the proposed art objects, the modification may not be artificial"—made by animal incisors or natural geologic forces rather than humans—"and on most it is not persuasively artistic" (*The Human Career*, 441).

5. In Africa south of the Zambezi, for example, forty-three sites have been excavated (not all professionally) in the last fifty years over an area of some 1.3 million square miles, one for every 30,000 square miles, whereas in southwestern France alone, with an area of some 12,000 square miles, more than a hundred sites have been explored, one for every 125 square miles.

6. The Australian record is fairly clear, but there is no good way of knowing if there were previous extinctions caused by Sapiens in Africa. Seven large mammalian genera did become extinct in Africa in the period after 130,000 years ago, but there is no way of knowing if they were hunted to extinction or were victims of climate change—either at the beginning of the Ice Age 71,000 years ago or at the end of it 10,000 years ago. The extinct genera were a variety of elephant, a variety of camel, three-toed horse, giant deer, large and medium hartebeests, and buffalo; Paul Martin, a specialist in extinctions at the University of Arizona, believes that the hartebeest extinctions at least "can be attributed to human hunters" (Martin and Klein, *Quaternary Extinctions*, 384).

Chapter 2: The Conquest of Europe

1. Claims have been made for sixteen Neandertal interments (La Chapelle-aux-Saintes, La Ferrassie, Le Roc de Marsal, Le Moustier, La Quina, Le Regourdou, La Roche à Pierrot, Feldhofer, Spy, Kiik-Koba, Teshik-Tash, Kebaara, Rabun, Amud, Dederiyeh, Shanidar) and two Sapiens interments (Qafez, Skhul) as burial sites in the period before late Sapiens burials after 30,000 years ago. Some show evidence that they were deliberately put in the ground, which is one thing, but as Richard Klein says of the Neandertal sites, they "tend to be very simple," as if used just as a means of disposal to prevent decay or carnivore interference, and there is "no clear indication of a burial ritual . . . or grave goods" (*The Human Career*, 551). Putative grave goods such as animal bones found in association with underground skeletons are most likely to be, as Klein says, a "fortuitous inclusion" of part of a common animal and there is no evidence at all of "burial ritual or ceremony."

2. Spencer Wells, in his *Journey of Man* (109), indicates that this was "the last substantial exchange between sub-Saharan Africa and Eurasia," since there are few genetic signs of any common markers after this date. As his work indicates, the population that settled Asia had a chromosome marker identical to one in those that settled the Levant, making them different from the remnant African population, but each of them also had different additional markers; the additional marker in the Levant population is identical to one in northeastern Africa (Ethiopia and Sudan in particular), proving that the Levant was settled by northeast Africans. The question of which route, Nile or Arabian, is not discernible, but there are no archeological remnants on the Arabian route.

3. Radiocarbon dating is a useful tool, but it has traditionally diminished in accuracy and effectiveness as one goes back much beyond about 30,000 years ago, so that a radiocarbon date of, say, 40,000 years ago is liable to be off—too "young"—by something around 4,000 years. New work by the distinguished Cambridge University archeologist Paul Mellars, however, has considerably sharpened the carbon-dating technique, protecting against the intrusion of more recent carbon and taking account of fluctuations in atmospheric carbon over the millennia. His study "A New Radiocarbon Revolution and the Dispersal of Modern Humans in Eurasia" (*Nature*, 23 February 2006) indicates that all the present carbon-dated dates for the period 50,000–30,000 years ago should be put back by as much as 4–5,000 years; for example, he has fixed the time of the Sapiens' initial dispersal across Europe to between 46,000 and 41,000 years ago (instead of 43,000–35,000 years). Similarly, work on the paintings of Chauvet Cave, once thought to date to 31,000 years ago, has now shown that it is more accurately put at 36–35,000 years (*Science*, 9 January 2004, 178–79, 202–7).

4. Darwin, in his discussions of sexual selection in human evolution (*The Descent of Man*, 1871), argued that "self-adornment" for "the admiration of others" was an instinct that evolved through sexual competition for mates. If so, it must have evolved fairly late, because there are no good signs of ornamentation or body painting in the record until after 100,000 years ago, and more convincingly after 70,000 years, although of course various bodily alterations like hair braiding, tattooing, scarification, and nose piercing—as various tribal people in the ethnographic record have done—would not fossilize. An argument has been made that in place of personal decoration, tools might have been a signal of fitness in earlier species—"a handaxe is a measure of strength, skill and character," Marek Kohn has said in *As We Know It* (London: Granta, 1999), so that may have been the *Homo erectus* way of sexual selection. Though since all handaxes look pretty much alike, to contemporary eyes at least, an Erectus female might not be able to distinguish much of a difference among her suitors that way.

5. There is some evidence for burials in Australia at a date that may be earlier than any in Europe, though it is not very robust and certainly not conclusive. Human remains found at Lake Mungo in New South Wales were covered with red ocher, possibly in-

dicating a burial ritual, and James M. Bowler of the University of Melbourne and his team have dated it to 40,000 (± 2,000) years ago (*Nature*, 20 February 2000). They have also found there a body they take to be cremated, of the same age.

Chapter 3: Intensification and Agriculture

1. Most skull injuries of Neandertals that are purported to be deliberately caused do not hold up to scrutiny, though one recently discovered skull of a 36,000-year-old Neandertal in Switzerland shows a fracture caused by a sharp weapon of some kind, which its discoverer, Christoph Zollikofer of the University of Zurich, argues is evidence of armed conflict. That the fracture healed while the man was still alive, though, shows that he was not in fact murdered, and indeed that he was taken care of long enough for the wound to heal. And a Neandertal skeleton from Shanidar shows evidence of a stab wound that punctured the lung, the eventual cause of death, but there is no way of assessing whether it was deliberate, accidental, or self-inflicted.

2. The giant deer, woolly rhinoceros, straight-toothed elephant, sabertooth cat, and woolly mammoth were extirpated entirely; other genera that disappeared and migrated elsewhere or else evolved into existing species include the giant rhinoceros, horse, cave bear, cave lion, primitive bison, dhole wolf, mountain goat, hippopotamus, musk ox, hyena, saiga antelope, hare, otter, vole, and jerboa (see Martin and Klein, *Quaternary Extinctions*, 384–85).

3. Vine Deloria Jr., in *Red Earth, White Lies*, argues that the passage over the Bering Strait would have been too difficult for either animals or humans to cross and the corridor from Alaska to the lower continent too inhospitable for migration that way. In the light of the established Sapiens sites along much of the Bering route, and the Edmonton site at the end of the passage, this does not hold much weight. His advocacy of a water passage for a number of Indian tribes entering North America is much better based, and it is plausible that Asians (like Australians) would have had boats capable of fairly long distances, if hugging the shorelines; no very great evidence for this exists, however. The idea that Indians could have come from anywhere but Asia has been disproved by Spencer Wells's evidence that Y chromosome markers of American natives match with those of Siberian populations, and suggest an entry to the Americas after 20,000 years ago.

4. The North American extinct genera were the giant armadillo, four species of ground sloths, two of bear, sabertooth and scimitartooth cats, cheetah, giant beaver, two species of capybara, horses, asses, onager, tapir, two species of peccary, camel, two species of llama, five of deer, stagmoose, saiga, shrub-ox, two species of musk ox, yak, mastodon, gomphothere, mammoth, dhole wolf, and short-faced skunk (see Martin and Klein, *Quaternary Extinctions*, 361–63).

5. It has recently been suggested that the widespread clearing of land for agriculture and

livestock, and extensive systems of irrigation (especially for rice), released great amounts of carbon dioxide and methane—the so-called greenhouse gases—into the atmosphere, starting about 8,000 years ago. William Ruddiman of the University of Virginia, in the 10 December 2003 issue of *Nature*, argues that these gases would have otherwise declined, as in earlier periods, but that with agriculture, "Humans were doing things on a scale that can explain why the natural trends failed."

Chapter 4: The Erectus Alternative

1. I am here conflating into *Homo erectus* a species that some paleologists choose to divide into *Homo ergaster* (or "African *Homo erectus*"), *Homo heidelbergensis, Homo antecessor,* and *Homo erectus,* reserving the last taxon for humans in Asia. As remarked in the notes to chapter 1, there is really very little significant morphological change through these "species," though the brain cage does tend to get larger over time, and it is simpler to give all these species a single name; as the archeologist Brian Fagan puts it in his *Journey from Eden,* "It is very hard to draw a clear taxonomic boundary between Homo erectus and archaic Homo sapiens on the one hand, and between archaic and anatomically modern Homo sapiens on the other" (60). In fact, there are some paleologists who argue that the *Homo erectus* of 1.8 million years ago was actually an early form of *Homo sapiens,* taking the broad view that the species changed only in mostly small ways to the present. Moreover, most archeologists have given up on the idea of a separate *Homo ergaster* species in Africa differing from the Asian *Homo erectus* species since the discovery in 2002 of a million-year-old skull from the Bouri Formation in Ethiopia with features like those of the Asian Erectus, proving that all were one species: "This fossil," according to the paleoanthropologist Tim White of the University of California, "is a crucial piece of evidence showing that the splitting of *H. erectus* into two species is not justified" (*Science,* 22 March 2002, 2193).

2. There is a whole range of estimates of total population during the middle Stone Age, from 125,000 and 500,000 to 1.3 and 1.5 million. The upper level seems reasonable to me: assuming, along with Shahin Rouhani of University College, London (in Mellars and Stringer, eds., *The Human Revolution,* 49), a total of 10,000 tribes in Africa and Asia and a tribal size of 150, that comes to 1.5 million and a population density of about 1 person per 10 square miles, about the density of modern hunter-gatherer tribes. Modern tribal sizes tend to cluster around 175 (100–200), according to Clive Gamble (in *The Paleolithic Settlement of Europe,* 50–51), often with larger "connubial" gatherings of 500, but my feeling is that the Erectus tribes would be slightly smaller, clustering around the 150 number that the psychologist Robin Dunbar (*Grooming, Gossip, and the Evolution of Language*) has argued is the evolutionary norm, since it is around the maximum number for cooperative egalitarian groups.

3. The figure of 30–40 per cent of food intake coming from meat is the assessment of the

anthropologist Richard B. Lee in the classic *Man the Hunter* collection of 1968 and has been duplicated and accepted by many anthropologists. A more recent study, recorded by Robert L. Kelly in *The Foraging Spectrum*, agrees that meat makes up about 35 percent of the diet of modern hunter-gatherer societies by weight but supplies 50 percent of the caloric needs.

4. It is only right to add that the title of the conference, and subsequent book, is not only sexist but quite inaccurate, as in many ethnographic societies—among them the Mbuti, Inuit, and Tiwi and Agta of the south Pacific—women do a fair share of the hunting.

SOURCE NOTES

Epigraphs

Ehrenreich, 46.
Darwin, Modern Library Edition (1936), 431.

Introduction

2: Genesis 1:30; 2:9, 15; 9:2, 3.

3 domination: Overview, *Atlas of Population and Environment*, AAAS, 2000 (online at www.ourplanet.com/aaas); Peter M. Vitousek et al., "Human Domination of Earth's Ecosystems," *Science*, 25 July 1997, 494–99; Andrew Balmford et al., "Economic Reasons for Conserving Wild Nature," *Science*, 9 August 2002, 950–53; Living Planet Report (panda.org); *New York Times*, 20 August 2002; Jeffrey D. Sachs, "Sustainable Development," *Science*, 30 April 2004, 649.

3 heat energy: Robert Wright and Luke Flynn, "Space-Based Estimate of the Volcanic Heat Flux into the Atmosphere during 2001 and 2002," *Geology* 32, no.3 (March 2004): 189–92.

3 imperilment: Union of Concerned Scientists, Washington, 1992; Ecosystem Millennium Assessment, *Science*, 1 April 2005; World Resources Institute, news release, 30 March 2005; *New York Times*, 31 March 2005; Wilson, *The Future of Life*, 27; *Scientific American*, February 2002; Rees, *Our Final Hour*.

4 "with the help": Alison S. Brooks and Richard Potts, "New Discoveries in Paleoanthropology: What's New, What's True, and What's Important?," *AnthroNotes* 21, no. 2 (winter–spring 1999–2000); see also vol. 24, no.1 (spring 2003).

6 art origins: Gary Snyder, "The Fiftieth Millennium," *Resurgence*, January–February 1999, 12; Pfeiffer, *The Creative Explosion*, 121; Bogucki, *The Origins of Human Society*, 83; Jonathan Adams, "Global Land Environments since the Last Interglacial," www.esd.ornl.gov/projects/qen/nerc.html; Michael Jochim, "Late Pleistocene Refugia in Europe," Soffer, ed., *The Pleistocene Old World*, 320–24.

7 **extinctions:** Martin and Klein, eds., *Quaternary Extinctions*, 385–91; Wells, *The Journey of Man*, 129, 133–34.

7 **"on the scene":** Wells, *The Journey of Man*, 126.

Chapter One: The Dawn of Modern Culture

11 **Mt. Toba explosion:** Michael Rampino, abstract for Environmental Catastrophes and Recoveries in the Holocene, 29 August–2 September 2002, Brunel University (atlas-conferences.com/cgi-bin/abstract), and with Stephen Self, "Volcanic Winter and Accelerated Glaciation following the Toba Super-Eruption," *Nature*, 3 September 1992, 50–52, and "Climate-Volcanism Feedback and the Toba Eruption of 74,000 Years Ago," *Quaternary Research* 40, no. 3 (1993): 269–80; *Discover*, December 1998; Meng-Yang Lee et al., "First Toba Supereruption Revival," *Geology* 32, no. 1 (2004): 61–64; Stanley H. Ambrose, "Late Pleistocene Human Population Bottlenecks, Volcanic Winter, and Differentiation of Modern Humans," *Journal of Human Evolution* 34, no. 6 (June 1998): 623–51 (and www.bradshawfoundation.com/evolution).

12 **ash particles:** William I. Rose et al., "Small Particles in Volcanic Eruption Clouds," *American Journal of Science* 280 (1980): 671; Joel Achenbach, "Big Chill," *National Geographic* 207, no.3 (March 2005): 1.

12 **"anatomically modern humans":** Ian McDougall et al., "Stratigraphic Placement and Age of Modern Humans from Kibish, Ethiopia," *Nature*, 17 February 2005, 733–36; Klein, *The Human Career*, 398–99, 586; Ann Gibbons, "Oldest Members of Homo Sapiens Discovered in Africa," *Science*, 13 June 2003, 1641.

13 **Gould's theory of punctuated equilibrium (with Niles Eldredge):** originally in T. Schopf, ed., *Models in Paleobiology* (San Francisco: W. H. Freeman, 1972); also, e.g., Gould, *Hen's Teeth and Horse's Toes* (New York: W. W. Norton, 1984), 259–60.

13 **gradual development:** Günter Bräuer, "The Evolution of Modern Humans: A Comparison of the African and Non-African Evidence," Mellars and Stringer, eds., *The Human Revolution*; Bräuer, "A Craniological Approach to the Origin of Anatomically Modern Homo Sapiens in Africa and Implications for the Appearance of Modern Europeans," Fred Smith and Frank Spencer, eds., *The Origins of Modern Humans* (New York: Alan R. Liss, 1984); G. Philip Rightmire, "Homo Sapiens in Sub-Saharan Africa," Smith and Spencer, eds., *The Origins of Modern Humans*; Robert Foley, "The Ecological Conditions of Speciation: A Comparative Approach to the Origins of Anatomically-Modern Humans," Mellars and Stringer, eds., *The Human Revolution*, 299–301; Fagan, *The Journey from Eden*, 58–61.

13 **brain size:** Klein, *The Human Career*, 145, 276, 377, 498; Gamble, *Timewalkers*, 147.

13 **Katanda:** Alison S. Brooks et al., "Dating the Context of Three Middle Stone Age Sites with Bone Points in the Upper Semliki Valley," *Science*, 28 April 1995, 548–53; John E. Yellen et al., "A Middle Stone Age Worked Bone Industry from Katanda, Upper Semliki

Valley, Zaire," *Science*, 28 April 1995, 553–56; McBrearty and Brooks, "The Revolution That Wasn't"; Klein, *The Human Career*, 439. (Yellen gives a date of 80,000 years ago with a 25 per cent chance of error, so something close to 70,000 is quite possible.)

13 **Zambia sites:** Lawrence S. Barham, "The Mumbwa Caves Project," *Nyame Akuma*, June 1995, 66–72; Lawrence S. Barham et al., "Bone Tools from Broken Hill (Kabwe) Cave, Zambia, and Their Evolutionary Significance," *Before Farming* 2 (2002).

13 **Tanzania sites:** McBrearty and Brooks, "The Revolution That Wasn't"; Constance Holden, "No Last Word on Language Origins," *Science*, 20 November 1998, 1455–58.

14 **Klein:** *The Human Career*, 440.

14 **Foley:** Mellars and Stringer, eds., *The Human Revolution*, 315.

14 **Southern Africa:** Klein, *The Human Career*, 401; Peter Mitchell, "Hunter-Gatherer Archaeology in Southern Africa: Recent Research, Future Trends," *Before Farming* 1 (2002); H. J. Deacon, "Late Pleistocene Palaeoecology and Archaeology in the Southern Cape, South Africa," Mellars and Stringer, eds., *The Human Revolution*; Deacon, Colloquium X, XIII International Union of Prehistoric and Protohistoric Sciences Congress, Italy, September 1996 (sun.ac.za/archeology/x-iupps); Deacon and Deacon, *Human Beginnings in South Africa*; Stanley H. Ambrose and Karl G. Lorenz, "Social and Ecological Models for the Middle Stone Age in Southern Africa," Mellars, ed., *The Emergence of Modern Humans*; *New York Times*, 26 February 2002.

15 **Cape plant life:** Jerold M. Lowenstein, "Counterpoints in Science," *California Wild* 52, no. 1 (winter 1999).

15 **Blombos Cave:** Christopher Henshilwood et al., "Blombos Cave, Southern Cape, South Africa: Preliminary Report on the 1992–1999 Excavations of the Middle Stone Age Levels," *Journal of Archaeological Science* 28, no. 4 (April 2001): 421–48; Christopher Henshilwood et al., "An Early Bone Tool Industry from the Middle Stone Age at Blombos Cave, South Africa: Implications for the Origins of Modern Human Behaviour, Symbolism and Language," *Journal of Human Evolution* 41, no. 6 (December 2001): 631–78; South African Museum, *MuseNews* 13, no. 4 (April 1999); Frederick Grine et al., "Human Remains from Blombos Cave, South Africa: (1997–1998 Excavations)," *Journal of Human Evolution* 38, no. 6 (June 2000): 755–65; Danika Painter, "Old Bone Tools Reveal Sharper Image of History," *ASU Research*, summer 2002, 24–27; Cape Field School Blombos Cave Project, naples.cc.sunysb.edu/CAS/cape.nsf/pages/Blombos; Deacon and Deacon, *Human Beginnings in South Africa*, 90, 93, 100, 109; *New York Times*, 2 December 2001.

16 **spear points:** Deacon and Deacon, *Human Beginnings in South Africa*, 100–101; Deacon, "Late Pleistocene Palaeoecology and Archaeology in the Southern Cape, South Africa," 557–61; Klein, *The Human Career*, 437–38.

17 **long-distance procurement:** Ambrose and Lorenz, "Social and Ecological Models for the Middle Stone Age in Southern Africa," 20–26.

17 **"marks the first":** Ambrose and Lorenz, "Social and Ecological Models for the Middle Stone Age in Southern Africa," 27; Cape Field School Blombos Cave Project, naples.cc .sunysb.edu/CAS/cape.nsf/pages/Blombos.

18 **"a precocious emergence":** Klein, *The Human Career*, 438, citing Paul Mellars, "Technological Changes at the Middle-Upper Palaeolithic Transition: Economics, Social Change and Cognitive Perspectives," Mellars and Stringer, eds., *The Human Revolution*.

18 **"bone tool industry":** Henshilwood et al., "An Early Bone Tool Industry from the Middle Stone Age at Blombos Cave, South Africa."

19 **animal bones:** Klein, *The Human Career*, 454–67; Ambrose and Lorenz, "Social and Ecological Models for the Middle Stone Age in Southern Africa," 7.

18 **animal weights:** Steven E. Churchill, "Weapon Technology, Prey Size Selection, and Hunting Methods in Modern Hunter-Gatherers: Implications for Hunting in the Palaeolithic and Mesolithic," Peterkin, Bricker, and Mellars, eds., *Hunting and Animal Exploitation in the Later Palaeolithic and Mesolithic of Eurasia*, 14–15; Elaine Anderson, "Who's Who in the Pleistocene: A Mammalian Bestiary," Martin and Klein, eds., *Quaternary Extinctions*, 80.

19 **hunted, not scavenged:** Curtis Marean et al., "Middle Stone Age Stratigraphy and Excavations at Die Kelders Cave 1 (Western Cape Province, South Africa): The 1992, 1993, and 1995 Field Seasons," *Journal of Human Evolution* 38, no. 1 (January 2000): 7–42; Curtis Marean, "A Critique of the Evidence for Scavenging by Neandertals and Early Modern Humans: New Data from Kobeh Cave (Zagros Mountains, Iran) and Die Kelders Cave 1 Layer 10 (South Africa)," *Journal of Human Evolution* 35, no. 2 (August 1998): 111–36; Klein, *The Human Career*, 459–63; Deacon and Deacon, *Human Beginnings in South Africa*, 76, 100; Cartmill, *A View to a Death in the Morning*, 16–17; Robert Blumenschine, *Early Hominid Scavenging Opportunities: Implications of Carcass Availability in the Serengeti and Ngorongoro Ecosystems* (Oxford: B.A.R., 1986).

19 **Klein:** Richard G. Klein, "Mammalian Extinctions and Stone Age People in Africa," Martin and Klein, eds., *Quaternary Extinctions*, 562; Richard G. Milo, "Evidence for Hominid Predation at Klasies River Mouth, South Africa, and Its Implications for the Behaviour of Early Modern Humans," *Journal of Archaeological Science* 25, no. 2 (February 1998): 99–133.

19 **aquatic species:** Robert C. Walter et al., "Early Human Occupation of the Red Sea Coast of Eritrea during the Last Interglacial," *Nature*, 4 May 2000, 65–69; McBrearty and Brooks, "The Revolution That Wasn't"; John Parkington, *Humanity from African Naissance to Coming Millennia: Colloquia in Human Biology and Palaeoanthropology* (Firenze: Firenze University Press, 2001).

19 **polyunsaturated fatty acids:** Parkington, *Humanity from African Naissance to Coming*

Millennia; Walter et al., "Early Human Occupation of the Red Sea Coast of Eritrea during the Last Interglacial"; Michael A. Crawford and David Arsch, *The Driving Force: Food, Evolution, and the Future* (New York: Harper and Row, 1989).

20 purposeful hunting: see e.g. Richard D. Alexander, "Evolution of the Human Psyche," Mellars and Stringer, eds., *The Human Revolution*, esp. 470; Cartmill, *A View to a Death in the Morning*, 230–31, 239; "The Nation," *New York Times*, 22 June 2003, § 4, 3, reported that only 6 per cent of Americans over sixteen are hunters.

20 Cartmill: Cartmill, *A View to a Death in the Morning*, 13.

21 Ehrenreich: *Blood Rites*, 46.

21 Erectus scavenging: R. J. Blumenschine, *Early Hominid Scavenging Opportunities*, and "Carcass Consumption Sequences and the Archaeological Distinction of Scavenging and Hunting," *Journal of Human Evolution* 15 (1986): 639–59; Cartmill, *A View to a Death in the Morning*, 1–27 (a demolition of the "human as killer ape" theory that was advanced by Raymond Dart and Robert Ardrey in *The Hunting Hypothesis* (New York: Atheneum, 1976), and others from the 1940s through the 1970s, showing that early Homo was not a hunter); Richard Klein, "Reconstructing How Early People Exploited Animals: Problems and Prospects," Matthew H. Nitecki and Doris V. Nitecki, eds., *The Evolution of Human Hunting* (New York: Plenum, 1987), 11–46; Margaret E. Lewis, "Carnivoran Paleoguilds of Africa: Implications for Hominid Food Procurement Strategies," *Journal of Human Evolution* 32, nos. 2–3 (February–March 1997): 257–88; Klein, *The Human Career*, 357–58, 459–63.

21 Dminisi: Klein, *The Human Career*, 316–19; Abesalom Vekua et al., "A New Skull of Early Homo from Dmanisi, Georgia," *Science*, 5 July 2002, 85–89.

21 Donggutuo: Klein, *The Human Career*, 271.

21 fire: Ralph Rowlett et al., discoveringarcheology.com/0599toc/5feature3=fire.shtml.

21 Neandertal hunting: see esp. Steven L. Kuhn, "Mousterian Technology as Adaptive Response: A Case Study," Peterkin, Bricker, and Mellars, eds., *Hunting and Animal Exploitation*; Nicolas Rolland, "Middle Palaeolithic Socio-Economic Formations in Western Eurasia: an Exploratory Survey," Mellars, ed., *The Emergence of Modern Humans*, esp. 353–54; and Patricia Anderson-Gerfaud, "Aspects of Behaviour in the Middle Palaeolithic: Functional Analysis of Stone Tools from Southwest France," Mellars, ed., *The Emergence of Modern Humans*; John J. Shea, "Neandertal and Early Modern Human Behavioral Variability," Fox, ed., *The Neanderthal Problem and the Evolution of Human Behavior*; Curtis Marean and Soo Yeun Kim, "Mousterian Large-Mammal Remains from Kobeh Cave," Fox, ed., *The Neanderthal Problem and the Evolution of Human Behavior*; Tattersall, *The Last Neanderthal*, 151–54; Stringer and Gamble, *In Search of the Neanderthals*, esp. 162–63; Klein and Edgar, *The Dawn of Human Culture*, 157–61; Paola Villa and Francesco d'Errico, "Bone and Ivory Points in the Lower and Middle Paleolithic of Europe," *Journal of Human Evolution* 41, no. 2 (August 2001): 69–112;

Lewis Binford, cited in Shreeve, *The Neandertal Enigma*, 159–63; Klein, *The Human Career*, 359–60, 459–63.

21 Schoningen: H. Thieme, "Altpaläolithische Holzgeräte aus Schöningen, Landkreis Helmstedt: Bedeutsame Funde zur Kulturentwicklung des frühen Menschen," *Germania* 77 (1999): 451–87.

23 hafting: Bruce L. Hardy et al., "Stone Tool Function at the Paleolithic Sites of Starosele and Buran Kaya III, Crimea: Behavioral Implications," *Proceedings of the National Academy of Science* 98, no. 19 (11 September 2001): 10972–77; Anderson-Gerfaud, "Aspects of Behaviour in the Middle Palaeolithic," 396–97; Paul Pettitt, "Odd Man Out: Neanderthals and Modern Humans," *British Archaeology* 51 (February 2000).

23 Vindija: Tom Higham et al., "Revised Direct Radiocarbon Dating of the Vindija G_1 Upper Paleolithic Neandertals," *Proceedings of the National Academy of Sciences* 103, no.3 (17 January 2006): 553–57.

24 clothes: Ralf Kittler et al., "Molecular Evolution of Pediculus Humanus and the Origin of Clothing," *Current Biology* 13, no. 16 (August 2003): 1414–17; *New York Times*, 19 August 2003.

25 "farming with fire": Deacon and Deacon, *Human Beginnings in South Africa*, 98–99; Deacon, "Late Pleistocene Palaeoecology and Archaeology in the Southern Cape, South Africa"; Deacon, "Landscapes of Exile and Healing: Climate and Gardens on Robben Island," *South African Archaeological Bulletin* 55, no. 172 (2000): 147–54; Shreeve, *The Neandertal Enigma*, 217.

25 Blombos art: Christopher Henshilwood et al., "Emergence of Modern Human Behavior: Middle Stone Age Engravings from South Africa," *Science*, 15 February 2002, 1278–80; Henshilwood et al., "Blombos Cave, Southern Cape, South Africa."

25 "art": Stanley Ambrose, news release, *Science*, 23 February 2002.

25 "We don't know": news release, BBC, 10 January 2002 (news.bbc.co.uk/hi/english/sci/tech/newsid_1753000/1753326).

25 Klein: Klein and Edgar, *The Dawn of Human Culture*, 14.

25 symbolic thinking: Mary Lecron Foster, "Symbolic Origins and Transitions in the Palaeolithic," Mellars, ed., *The Emergence of Modern Humans*; Terrence Deacon, *The Symbolic Species* (New York: W. W. Norton, 1997).

27 language: Walker and Shipman, *The Wisdom of the Bones*, 210–28; Stringer and Gamble, *In Search of the Neanderthals*, 88–91; Gamble, *Timewalkers*, 170–75; Philip Lieberman, "The Origins of Some Aspects of Human Language and Cognition," Mellars and Stringer, eds., *The Human Revolution*; "Evolution of Language," *Science*, 27 February 2004 [special issue]; Tattersall, *Becoming Human*; Paul Mellars and Kathleen Gibson, *Modelling the Early Human Mind* (Cambridge: McDonald Institute, 1996); *Cambridge Archaeological Journal* 8, no. 1 (April 1998) [special issue]; Larry Trask et al., "The Origins of Human Speech," *Cambridge Archaeological Journal*, April

1998:69–94; Holden, "No Last Word on Language Origins"; *New York Times*, 15 July 2003.

27 Mellars: Mellars, "Technological Changes at the Middle-Upper Palaeolithic Transition," 354–60.

27 Tattersall: Tattersall, *The Monkey in the Mirror*, 152–53.

27 speech gene: Nicholas Wade, "Language Gene Is Traced to Emergence of Humans," *New York Times*, 15 August 2002, § A, 18; Wade, "Researchers Say Gene Is Linked to Language," *New York Times*, 4 October 2001, § A, 1; *Science*, 16 August 2002; Klein and Edgar, *The Dawn of Human Culture*, 271.

27–28 mammal bone: Francesco d'Errico, "An Engraved Bone Fragment from c. 70,000-Year-Old Middle Stone Age Levels at Blombos Cave, South Africa: Implications for the Origin of Symbolism and Language," *Antiquity*, June 2001, 308–17.

28 early decorations: McBrearty and Brooks, "The Revolution That Wasn't," 528.

28 ocher fragments: Peter Mitchell, "Hunter-Gatherer Archeology in Southern Africa," *Before Farming* 1 (2002): 5; McBrearty and Brooks, "The Revolution That Wasn't," 522–25; Deacon, "Late Pleistocene Palaeoecology and Archaeology in the Southern Cape, South Africa," 559.

28 ornamentation: Christopher Henshilwood et al., "Middle Stone Age Shell Beads from South Africa," *Science*, 16 April 2004, 404 (where he gives a probable date of 75.6 million years, with a margin of error of 3.4). I discount the recent shells found at Skhul, said to be 100,000 years old or so, as poorly dated and unconvincing as ornaments.

29 60,000 years ago: Deacon and Deacon, *Human Beginnings in South Africa*, 106.

29 Deacon: Deacon, "Late Pleistocence Palaeoecology and Archaeology in the Southern Cape, South Africa," 558–91, esp. 561.

29 Howieson's Poort: Deacon and Deacon, *Human Beginnings in South Africa*, 106; Deacon, "Late Pleistocene Palaeoecology and Archaeology in the Southern Cape, South Africa."

30 longer life: Klein, *The Human Career*, 554.

30 arrows: McBrearty and Brooks, "The Revolution That Wasn't," 502.

30 McBrearty and Brooks: McBrearty and Brooks, "The Revolution That Wasn't," 532; Clark, "The Origins and Spread of Modern Humans," also cited in Cohen, *The Food Crisis in Prehistory*, 102.

30 nucleotides: S. T. Sherry et al., "Mismatch Distributions of mtDNA Reveal Recent Human Population Expansions," *Human Biology* 66, no. 5 (October 1994): 761–75; Shreeve, *The Neandertal Enigma*, 122.

30 Greenland ice core: See e.g. Greenland Ice Sheet Project Two, "Pre-Holocene Rapid Climate Change" (gust.sr.unh.edu/GISP2/Data).

31 migrating herds: Robert Foley and Marta Mirazón Lahr, "Mode 3 Technologies and the Evolution of Modern Humans," *Cambridge Archaeological Journal* 7, no. 1 (April 1997): 3–36, and antiquityofman.com/mode3technologies.html.

31 East Africa: McBrearty and Brooks, "The Revolution That Wasn't," esp. 521–24.

31 Enkapune Ya Moto: Stanley Ambrose, "Chronology of the Later Stone Age and Food Production in East Africa," *Journal of Archaeological Science* 25, no. 4 (April 1998): 377–92; Klein and Edgar, *The Dawn of Human Culture*, 11–16; Stanley H. Ambrose, "Small Things Remembered: Origins of Early Microlithic Industries in Sub-Saharan Africa," Robert B. Elston and Steven L. Kohn, eds., *Thinking Small: Global Perspectives on Microlithization*, Archeological Papers of the American Anthropology Association 12 (2002).

32 spear thrower: Malcolm F. Farmer, "The Origins of Weapons Systems," *Current Anthropology* 35, no. 5 (December 1994): 679–81.

32 Aterian: Philip Van Peer, "Nile Corridor and the Out-of-Africa Model," Fox, *The Neanderthal Problem and the Evolution of Human Behavior*, esp. 122–23, 129, and see 132; Klein, *The Human Career*, 434–36 (his dates for this industry are later); Gertrude Caton-Thompson, *The Aterian Industry* (London: Royal Anthropological Institute, 1946); Foley and Lahr, "Mode 3 Technologies and the Evolution of Modern Humans"; Mauro Cremaschi, "Some Insights on the Aterian in the Libyan Sahara: Chronology, Environment, and Archaeology," *African Archaeological Review* 15, no. 4 (1998): 261–86.

32 "the origin": Caton-Thompson, as quoted in Van Peer, "Nile Corridor and the Out-of-Africa Model."

32 Lupemban: McBrearty and Brooks, "The Revolution That Wasn't."

33 migrations: Foley and Lahr, "Mode 3 Technologies and the Evolution of Modern Humans"; Klein, *The Human Career*, 314–27; *Science*, 2 March 2001 [special issue]; Alan Templeton, "Out of Africa Again and Again," *Nature*, 7 March 2002, 45–51; Peter Forster and Shuichi Matsumura, "Did Early Humans Go North or South," *Science*, 13 May 2005, 965–66; Vincent Macaulay et al., "Single Rapid Coastal Settlement of Asia Revealed by Analysis of Complete Mitochondrial Genomes," *Science*, 13 May 2005, 1034–36.

33 Dmanisi: Josh Fischman, "Family Ties," *National Geographic*, April 2005, 18–27; Klein, *The Human Career*, 316–18, though he favors a date of about 1 million years.

33 Asian Erectus: Rightmire, *The Evolution of Homo Erectus*, 12–14; Klein, *The Human Career*, 328–33.

33 Nihewan Basin: R. X. Zhu et al., "New Evidence on the Earliest Human Presence at High Northern Latitudes in Northeast Asia," *Nature*, 30 September 2004, 559–62.

33 Gibraltar: Darren A. Fa et al., gibraltar.gi/museum.

33 Sapiens migrations to Asia: Macaulay et al., "Single Rapid Coastal Settlement of Asia"; Klein, *The Human Career*, 489–91, 498, 566–72; Wells, *The Journey of Man*, 108, 119–21, 75–80; Rhys Jones, "East of Wallace's Line: Issues and Problems in the Colonization of the Australian Continent," Mellars and Stringer, eds., *The Human Revolution*; Michael D. Petraglia, "The Lower Paleolithic of the Arabian Peninsula: Occupations, Adaptation, and Dispersals," *Journal of World Prehistory* 17, no. 2 (June 2003): 141–79; Martha Lahr and Robert Foley, "Multiple Dispersals and Modern Human Origins," *Evolutionary Anthropology* 3, no.2 (1994): 48–60; Christophe Coupé and J. M. Hombert, "From Africa to Australia," ddl.ish-lyon.cnrs.fr/biblio. Additional genetic data for these migrations: Wells, *The Journey of Man*, 73–80; Mary-Claire King and Arno G. Motulsky, "Mapping Human History," *Science*, 20 December 2002, 2342–43; Lluis Quintana-Murci et al., "Genetic Evidence of an Early Exit of Homo Sapiens from Africa through Eastern Africa," *Nature Genetics* 23, no. 4 (December 1999): 437–41; Nicholas Wade, "At Genetic Frontier, the House Mouse Serves Humanity," *New York Times*, 10 December 2002, § F, 3.

33 Bab el Mandeb: Christopher Stringer, "Coasting Out of Africa," *Nature*, 14 May 2000, 24–26.

34 Niah Cave: Graeme Barker, "Prehistoric Foragers and Farmers in South-East Asia: Renewed Investigations at Niah Cave, Sarawak," *Proceedings of the Prehistoric Society* 68 (2002): 147–64.

34 "a fundamental change": Mellars and Stringer, eds., *The Human Revolution*, 752.

34 "the time required": Coupe and Hombert, "From Africa to Australia," 10, citing Luigi Cavalli-Sforza.

34 genetic studies: Yuehai Ke et al., "African Origin of Modern Humans in East Asia: A Tale of 12,000 Y Chromosomes," *Science*, 11 May 2001, 1151–53 (China); ABC.net.au .science/news/stories, 5 November 2001 (China); Quintana-Murci et al., "Genetic Evidence of an Early Exit of Homo Sapiens from Africa through Eastern Africa" (India); Max Ingman et al., "Mitochondrial Genome Variation and the Origin of Modern Humans," *Nature*, 7 December 2000, 708–13; Nicholas Wade, "Genes Tell New Story on the Spread of Man," *New York Times*, 7 December 1999, § F, 2; *Time Asia*, 17 January 2000; dispersal at 50,000 years also figured by Carol Lalueza Fox, quoted in Madhusree Mukerjee, "Out of Africa, into Asia," *Scientific American*, January 1999, 24; Macaulay et al., "Single Rapid Coastal Settlement of Asia," gives 85,000–55,000, median 70,000; Sandra Bowdler, "The Pleistocene Pacific," Donald Denoon, ed., *The Cambridge History of the Pacific Islanders* (Cambridge: Cambridge University Press, 1997) (Australia); Klein, *The Human Career*, 566–72, esp. 566–67, questioning dates; Sandra Bowdler, "Peopling Australasia: The 'Coastal Colonization,'" Mellars, ed., *The Emergence of Modern Humans*.

34 Devil's Lair: Chris S. M. Turney et al., "Early Human Occupation at Devil's Lair, Southwestern Australia 50,000 Years Ago," *Quaternary Research* 55 (2001): 3–13.

34 Lake Mungo: Forster and Matsumura, "Did Early Humans Go North or South."

34 Australian dates: James M. Bowler et al., "New Ages for Human Occupation and Climatic Change at Lake Mungo, Australia," *Nature*, 20 February 2003, 837–40; *Science*, 24 October 2003; Nicholas Wade, "Dating of Australian Remains Backs Theory of Early Migration of Humans," *New York Times*, 19 February 2003, § A, 6; Peter Brown, "The First Australians: The Debate Continues," *Australasian Science* 21 (May 2000): 28–31, and "Australian and Asian Palaeoanthropology," personal.une.edu.au/pbrown3/palaeo.html.

34 Malakunanja II, Nauwalabila I: Klein, *The Human Career*, 566–68.

35 Erectus survival dates: Klein, *The Human Career*, 273.

35 extinctions: Richard G. Roberts et al., "New Ages for the Last Australian Megafauna: Continent-wide Extinction about 46,000 Years Ago," *Science*, 8 June 2001, 1888–92; Paul S. Martin, "Prehistoric Overkill: The Global Model," Martin and Klein, eds., *Quaternary Extinctions*, 358, 376–80 (date 40,000 years, 380), and, in the same source, nine other articles on Australia, esp. Duncan Merrilees, "Comings and Goings of Late Quaternary Mammals near Perth in Extreme Southwestern Australia," 630, David R. Horton, "Red Kangaroos: Last of the Australian Megafauna," 673, and Jared M. Diamond, "Historic Extinction: A Rosetta Stone for Understanding Prehistoric Extinctions," 838.

35 climate: Gifford Miller et al., "Pleistocene Extinction of Genyornis Newtoni: Human Impact on Australian Megafauna," *Science*, 8 January 1999, 205–8.

36 New Guinea: Gamble, *Timewalkers*, 222.

36 Eldredge: Eldredge, *Dominion*, 82.

Chapter Two: The Conquest of Europe

37 band size: see e.g. Kingdon, *Self-Made Man*, 285.

37 Nile corridor: Fagan, *The Journey from Eden*, 71–72; Wells, *The Journey of Man*, 108–10; Philip Van Peer, "Nile Corridor and the Out-of-Africa Model"; Fox, *The Neanderthal Problem and the Evolution of Human Behavior*; Van Peer and Pierre M. Vermeersch, "Middle to Upper Palaeolithic Transition: The Evidence for the Nile Valley," Mellars, ed., *The Emergence of Modern Humans*; Ofer Bar-Yosef, "The Middle and Early Upper Paleolithic in Southwest Asia and Neighboring Regions," Bar-Yosef and David Pilbeam, eds., *The Geography of Neanderthals and Modern Humans in Europe and the Greater Mediterranean* (Cambridge, Mass.: Peabody Museum, 2000); Anthony E. Marks, "The Middle and Upper Palaeolithic of the Near East and the Nile Valley: The Problem of Cultural Transformations," Mellars, ed., *The Emergence of Modern Humans*, doubts Nile corridor.

38 hiving-off rate: Esmee Webb and David Rindos, "The Mode and Tempo of the Initial Human Colonization of Empty Landmasses: Sahul and the Americas Compared," C. Michael Barton and Geoffrey A. Clark, eds., *Rediscovering Darwin*, Archeological Papers of the American Anthropological Association 7 (1997): 237.

38 Rolland: *Current Anthropology* 39 (1998): suppl. S133–34.

38 Nubian culture: Van Peer, "Nile Corridor and the Out-of-Africa Model."

38 "hunting": Van Peer, "Nile Corridor and the Out-of-Africa Model,"120, 129.

39 "a context": Van Peer, "Nile Corridor and the Out-of-Africa Model,"130.

39 Taramsa skeleton: Pierre M. Vermeersch et al., "A Middle Palaeolithic Burial of a Modern Human at Taramsa Hill, Egypt," *Antiquity* 72, no. 277 (September 1998): 475–84 (gives dates of 80,400 to 49,800 years ago, with a mean of 55,000); B.B., "Ancient Child's Burial on the Nile," *Science News*, 10 October 1998, 235 (55,000); according to Stephen Stokes of the original team, work is ongoing to refine the dates (personal communication).

39 Levant: Ofer Bar-Yosef, "Symbolic Expressions in Later Prehistory of the Levant: Why Are They So Few?"; Conkey et al., eds., *Beyond Art*; Nitecki and Nitecki, eds., *Origins of Anatomically Modern Humans*; Soffer, ed., *The Pleistocene Old World*; G. A. Clark and J. M. Lindly, "The Case of Continuity: Observations on the Biocultural Transition in Europe and Western Asia," Mellars and Stringer, eds., *The Human Revolution*; Anthony E. Marks, "The Middle and Upper Palaeolithic of the Near East and the Nile Valley"; Anthony E. Marks, "The Early Upper Paleolithic: The View from the Levant"; Knecht et al., eds., *Before Lascaux*; K. Ohnuma and C. A. Bergman, "A Technological Analysis of the Upper Palaeolithic Levels (XXV–VI) of Ksar Akil, Lebanon," Mellars, ed., *The Emergence of Modern Humans*; Steven L. Kuhn et al., "Initial Upper Paleolithic in South-Central Turkey and Its Regional Context: A Preliminary Report," *Antiquity* 73 (1999): 505–17; Stanley H. Ambrose, "Implications of Geomagnetic Field Variation for the Chronology of the Middle Later Stone Age and Middle Upper Paleolithic Transitions," *Abstracts of the Paleoanthropology Society Meetings*, April 1997, 2.

40 Y-shaped patterns: Bar-Yosef, "The Middle and Early Upper Paleolithic in Southwest Asia and Neighboring Regions."

40 "an African origin": Klein, *The Human Career*, 489.

40 skeleton types: see e.g. Trenton W. Holliday, "Evolution at the Crossroads: Modern Human Emergence in Western Asia," *American Anthropologist* 102, no. 1 (March 2000): 54–68.

40 Y chromosomes: Wells, *The Journey of Man*, 109.

40 Boker Tachtit: Ofer Bar-Yosef, "The Contributions of Southwest Asia to the Study of the Origin of Modern Humans," Nitecki and Nitecki, eds., *Origins of Anatomically Modern Humans*, 40–43; Clark and Lindly, "The Case of Continuity," 651–53; Marks, "The Middle and Upper Palaeolithic of the Near East and the Nile Valley," 67–72; Sheldon Klein, "Human Cognitive Changes at the Middle to Upper Palaeolithic Transitition: The Evidence of Boker Tachtit," Mellars, ed., *The Emergence of Modern Humans*, 499–516.

40 blades: Klein, *The Human Career*, 520.

41 Sheldon Klein, "a new pattern," "analogical": Mellars, ed., *The Emergence of Modern Humans*, 499.

42 Ksar 'Akil: Marks, "The Middle and Upper Palaeolithic of the Near East and the Nile Valley," 65–74; Ohnuma and Bergman, "A Technological Analysis of the Upper Palaeolithic Levels (XXV–VI) of Ksar Akil, Lebanon"; Clark and Lindly, "The Case of Continuity"; Klein, *The Human Career*, 489.

42 art: Alexander Marshack, "Paleolithic Image Making and Symboling in Europe and the Middle East: A Comparative Review," Conkey et al., eds., *Beyond Art*, 60–64; Bar-Yosef, "Symbolic Expressions in Later Prehistory of the Levant," 167–69.

43–44 Freud, "omnipotence of thought": A. A. Brill, ed., *The Basic Writings of Sigmund Freud* (New York: Modern Library, 1938), 868, 873–76.

44 Frazer: Frazer, *The Golden Bough*, 49.

44 ornamentation: Steven Kuhn et al., "Ornaments of the Earliest Paleolithic: New Insights from the Levant," *Proceedings of the National Academy of Science* 98, no.13 (19 June 2001): 7641–46 (on Üçağizli); Steven Kuhn, *Gibraltar Conference*, August 2001, gib.gi/museum/p10.

44 Ksar 'Akil: Bogucki, *The Origins of Human Society*, 99.

45 other beads, bone point: Bar-Yosef, "Symbolic Expressions in Later Prehistory of the Levant," 166–67, 168.

45 "a shared system," "We might expect": Kuhn et al., "Ornaments of the Earliest Paleolithic."

45 Hayonim cave: Mary Stiner et al., "Paleolithic Population Growth Pulses Evidenced by Small Animal Exploitation," *Science*, 8 January 1999, 190–94.

46 Bar-Yosef: Nitecki and Nitecki, eds., *Origins of Anatomically Modern Humans*, 54–55.

46 migrations from the Levant: Gilbert Tostevin, "The Middle to Upper Paleolithic Transition from the Levant to Central Europe: In Situ Development or Diffusion?," Jorg Orschiedt and Gerd-Christian Weniger, eds., *Neanderthals and Modern Humans: Discussing the Transition: Central and Eastern Europe from 50,000–30,000 BP* (Mettmann, Germany: Neanderthal Museum, 2000), 92–111, esp. 94; Bar-Yosef and Pilbeam, eds., *The Geography of Neanderthals and Modern Humans in Europe and the Greater Mediterranean*, 7; Scott J. Brown, "Neanderthal and Modern Humans," www.Neanderthal-modern.com./ceeurope.

47 extent of settlement: Klein, *The Human Career*, 479, 484–86, 558–59.

47 climate: Gamble, *The Paleolithic Societies of Europe*, 280.

47 Georgia: Ofer Bar-Yosef et al., "Paleolithic Research in Western Transcaucasia," paper presented at the meeting of the Paleoanthropology Society, March 1998.

48 genetic evidence: Wells, *The Journey of Man*, 110–17, 132–34; Nicholas Wade, "The Origin of the Europeans," *New York Times*, 14 November 2000, § F, 9; Martin Richards et al., "In Search of Geographical Patterns in European Mitochondrial DNA," *American Journal of Human Genetics* 71 (2002): 1168–74.

48 "earliest migration": Wade, "The Origin of the Europeans."

48 "trace their ancestry": Wells, *The Journey of Man*, 111.

48 Bacho Kiro: Klein, *The Human Career*, 486; Brown, "Neanderthal and Modern Humans," 3; Gamble, *The Paleolithic Societies of Europe*, 277–78, 297.

48 Pestera cu Oase: Erik Trinkaus et al., "An Early Modern Human from the Pestera cu Oase, Romania," *Proceedings of the National Academy of Sciences* 100, no. 20 (30 September 2003): 11231–36; *Science*, 9 May 2003; Washington University, "Bones from French Cave Show Neanderthals, Cro-Magnon Hunted Same Prey," 22 September 2003 [news release] (eurekalert.org/pub_releases/2003-09/uow-bff092203.php).

48 Kostenki, Kent's Cavern: Gamble, *The Paleolithic Societies of Europe*, 280.

48 Abri Tapolca, Silicka Brzova: Brown, "Neanderthal and Modern Humans."

48 Cro Magnon: Klein, *The Human Career*, 477.

48–49 Temnata and Boker Tachtit: Gilbert Tostevin, "The Middle to Upper Paleolithic Transition from the Levant to Central Europe: In Situ Development or Diffusion?," Orschiedt and Weniger, eds., *Neanderthals and Modern Humans*, 90–109; Stranska skala and Boker Tachtit, "P. Skrdla," iabrno.cz.bt; also Janusz K. Kozlowski, "Cultural Context of the Last Neanderthals and Early Modern Humans in Central-Eastern Europe," Ofer Bar-Yosef et al., eds., *Lower and Middle Paleolithic* (Forli, Italy: Abard, 1996).

49 sites 46–40,000 years: Klein, *The Human Career*, 484–86; Gamble, *The Paleolithic Societies of Europe*, 275–78; and a list in Jean-Pierre Bocquet-Appel and Pierre Yves Demars, "Neanderthal Contraction and Modern Human Colonization of Europe," *Antiquity* 74, no. 3 (2000): 544–52.

49 Fumane: Steven Kuhn, "Pioneers of Microlithization: The 'Proto-Aurignacian' of Southern Europe," *Archeological Papers of the American Anthropological Association* 12, no. 1 (2002): 83–93.

49 L'Arbreda: Gamble, *The Paleolithic Societies of Europe*, 276–77.

49 Abri Pataud: Gamble, *The Paleolithic Societies of Europe*, 275.

49 Neandertals: Klein, *The Human Career*, 368 (range), 374–77 (dates).

50 extinction: Shreeve, *The Neandertal Enigma*, 342 (Spain); Higham et al., "Revised Direct Radiocarbon Dating of the Vindija G_1 Upper Paleolithic Neandertals" (Croatia); Tattersall, *The Last Neanderthal*, 198–203.

50 Pettitt: Paul Pettitt, "Odd Man Out: Neanderthals and Modern Humans," *British Archaeology* 51 (February 2000).

50 Shea: Joe Alper, "Rethinking Neanderthals," *Smithsonian* 34, no. 3 (June 2003): 83.

50 Neandertal skeletons: Klein, *The Human Career*, 474–75; Alper, "Rethinking Neanderthals"; Erik Trinkaus, "The Neandertals and Modern Human Origins," *Annual Review of Anthropology* 15 (1986): 193–218; Valerius Geist, "The Neanderthal Paradigm," cogweb.ucla.edu/ep/NeanderthalParadigm.html; Stringer and Gamble, *In Search of the Neanderthals*, 94.

51 mixed skeletons: Bräuer, "The Evolution of Modern Humans," 137–39; Gambier, "Fossil Hominids from the Early Upper Palaeolithic (Aurignacian) of France," 197; Ciddalia Duarte et al., "The Early Upper Paleolithic Human Skeleton from the Abrigo do Lagar Velho (Portugal) and Modern Human Emergence in Iberia," *Proceedings of the National Academy of Sciences* 96, no. 13 (22 June 1999): 7604–9, refuted by Ian Tattersall and Jeffrey H. Schwartz, "Hominids and Hybrids: The Place of Neanderthals in Human Evolution," *Proceedings of the National Academy of Sciences* 96, no. 13 (22 June 1999): 7117–19.

51 Klein: Klein, *The Human Career*, 482.

51 Neandertal diet: Michael Richards et al., "Stable Isotope Evidence for Increasing Dietary Breadth in the European Mid-Upper Paleolithic," *Proceedings of the National Academy of Sciences* 98, no. 11 (22 May 2001): 6528–32; Klein, *The Human Career*, 532.

51 Zubrow: Mellars and Stringer, eds., *The Human Revolution*, 212–31.

51 Eldredge: Eldredge, *Dominion*, 86.

52 Ice Age climate: Gamble, *The Paleolithic Settlement of Europe*, 186–87, 280; GISP data in Lowell Stott et al., "Super ENSO and Global Climate Oscillations at Millennial Time Scales," *Science*, 12 July 2002, 224.

52 mammals: Klein, *The Human Career*, 531; Gamble, *The Paleolithic Settlement of Europe*, 331–32.

53 Soffer: Soffer, "The Middle to Upper Palaeolithic Transition on the Russian Plain," Mellars and Stringer, eds., *The Human Revolution*, 733, 736.

53 Mellars, weaponry: Paul Mellars, "Technological Changes at the Middle-Upper Palaeolithic Transition: Economics, Social Change and Cognitive Perspectives," Mellars and Stringer, eds., *The Human Revolution*, 339–52.

53 tools: Klein, *The Human Career*, 537–38.

53 fur trapping: Klein, *The Human Career*, 535.

54 Bacho Kiro: Janusz K. Koslowski, "A Multiaspectual Approach to the Origins of the Upper Palaeolithic in Europe," Mellars, ed., *The Emergence of Modern Humans*, 430; Gamble, *The Paleolithic Settlement of Europe*, 331–32.

54 Dolní Vestonice: Shreeve, *The Neandertal Enigma*, 283, 289; Klein, *The Human Career*, 544.

54 reciprocity: Shreeve, *The Neandertal Enigma*, 299.

54 bone industry: Mellars, "Technological Changes at the Middle-Upper Palaeolithic Transition"; Francis B. Harrold, "Variability and Function among Gravette Points from Southwestern France," Peterkin, Bricker, and Mellars, eds., *Hunting and Animal Exploitation*; Clark and Lindly, "The Case of Continuity," 645; Heidi Knecht, "Early Upper Paleolithic Approaches to Bone and Antler Projectile Technology," Peterkin, Bricker, and Mellars, eds., *Hunting and Animal Exploitation*; Gail Larsen Peterkin, "Lithic and Organic Hunting Technology in the French Upper Palaeolithic," Peterkin, Bricker, and Mellars, eds., *Hunting and Animal Exploitation*.

55 light spearpoints: Anne Pike-Tay, "Hunting in the Upper Périgordian: A Matter of Strategy or Expedience?," Knecht et al., eds., *Before Lascaux*, 53.

55 V-shaped split, reindeer, Abri Blanchard, Isturitz: Knecht, "Early Upper Paleolithic Approaches to Bone and Antler Projectile Technology."

55 ornaments: Randall White, "Production Complexity and Standardization in Early Aurignacian Bead and Pendant Manufacture: Evolutionary Implications," Mellars and Stringer, eds., *The Human Revolution*; Randall White, "Personal Ornaments from the Grotte de Renne at Arcy-sur-Cure," *Athena Review* 2, no. 4 (2001): 41–46; White, "Technological and Social Dimensions of 'Aurignacian-Age' Body Ornaments across Europe," Knecht et al., eds., *Before Lascaux*; Randall White, "Substantial Acts: From Materials to Meaning in Upper Paleolithic Representation," Conkey et al., *Beyond Art*, 93–122, and Institute for Ice Age Studies (Insticeagestudies.com); Mellars, "Technological Changes at the Middle-Upper Palaeolithic Transition," 343.

55 Bacho Kiro: White, "Personal Ornaments from the Grotte de Renne at Arcy-sur-Cure."

56 "the explosion": White, "Production Complexity and Standardization in Early Aurignacian Bead and Pendant Manufacture," 367.

56 "a complex production": White, "Case Study I, Aurignacian Personal Ornaments, C," Insticeagestudies.com.

56 extensive distances: White, "Production Complexity and Standardization in Early Aurignacian Bead and Pendant Manufacture," 375–77; Klein, *The Human Career*, 544; Brooke S. Blades, "Aurignacian Lithic Economy and Early Modern Human Mobility: New Perspectives from Classic Sites in the Vézère Valley of France," *Journal of Human Evolution* 37, no. 1 (July 1999): 91–120.

56 Kostenki and Pavlov: Gamble, *Timewalkers*, 186–87; Gamble, *The Paleolithic Settlement of Europe*, 377.

57 art: Conkey et al., eds., *Beyond Art*; Pfeiffer, *The Creative Explosion*; Klein, *The Human Career*, 545–53; Alain Leroi-Gourhan, *The Dawn of European Art* (Cambridge: Cambridge University Press, 1982); Clottes and Courtin, *The Cave beneath the Sea*; Jean-Marie Chauvet et al., *Dawn of Art: The Chauvet Cave* (New York: Harry N. Abrams, 1996); White, Insticeagestudies.com/library/earliestimages; James Q. Jacobs, geocities .com/archaeogeo/paleo/dawn.

57 figurines: Alexander Marshack, "Early Hominid Symbol and Evolution of the Human Capacity," Mellars, ed., *The Emergence of Modern Humans*; Marshack, "Paleolithic Image Making and Symboling in Europe and the Middle East"; Alexander Marshack, "The Female Image: A 'Time Factored' Symbol: A Study in Style and Aspects of Image Use in the Upper Palaeolithic," *Proceedings of the Prehistoric Society* 57, no. 1 (1991): 17–31; White, "Substantial Acts"; Iain Davidson, "The Power of Pictures," Conkey et al., eds., *Beyond Art*, 125–60; Olga Soffer, "The Mutability of Upper Paleolithic Art in Central and Eastern Europe: Patterning and Significance," Conkey et al., eds., *Beyond Art*, 239–62; Margherita Mussi and Daniela Zampetti, "Carving, Printing, Engraving:

Problems with the Earliest Italian Design," Conkey et al., eds., *Beyond Art*, 217–38; Paul G. Bahn, "New Advances in the Field of Ice Age Art," Nitecki and Nitecki, eds., *Origins of Anatomically Modern Humans*; Olga Soffer, "Upper Paleolithic Connubia, Refugia, and the Archaeolgical Record from Eastern Europe," Soffer, ed., *The Pleistocene Old World*, 334–39; Gamble, *The Paleolithic Settlement of Europe*; Pfeiffer, *The Creative Explosion*, 202–6, 237–38; Klein, *The Human Career*, 480, 549–50; Chauvet et al., *Dawn of Art*, 124; Olga Soffer et al., "The 'Venus' Figurines: Textiles, Basketry, Gender, and Status in the Upper Paleolithic with CA Comment," *Current Anthropology* 41, no. 4 (August–October 2000): 511–38.

57 **"marked":** Marshack, "Early Hominid Symbol and Evolution of the Human Capacity," 478.

57 **"tens of thousands":** Pfeiffer, *The Creative Explosion*, 7.

58 **Hohlenstein-Stadel:** Marshack, "Early Hominid Symbol and Evolution of the Human Capacity," 478–80; Wolfgang M. Werner, "Lionman Home Page," home.bawue.de/~wmwerner/english/lionman. A second lionman was found at Hohle Fels Cave in 2002, dated to about 30,000 in Rex Dalton, "Lion Man Takes Pride of Place as Oldest Statue," *Nature*, 4 September 2003, 7.

59 **"Venus figurines":** Soffer, ed., *The Pleistocene Old World*; White, "Substantial Acts," 107–19; Klein, *The Human Career*, 550; Timothy Taylor, "Uncovering the Prehistory of Sex," *British Archaeology* 15 (June 1996); Joseph Campbell, *Primitive Mythology* (New York: Penguin, 1976), 313–34.

59 **Moravian figurines:** Gamble, *The Paleolithic Societies of Europe*, 403–4; Olga Soffer et al., "The Pyrotechnology of Performance Art: Moravian Venuses and Wolverines," Knecht et al., *Before Lascaux*.

59 **Soffer:** Shreeve, *The Neandertal Enigma*, 281.

59 **forty hours, "carried out by":** Paul Hahn, "Aurignacian Art in Central Europe," Knecht et al., eds., *Before Lascaux*, 272.

60 **350 caves, 15 per cent:** White, "The Earliest Images," Insticeagestudies.com/library.

60 **15,000 paintings:** Pfeiffer, *The Creative Explosion*, 2, though that is undoubtedly an underestimate.

60 **"The animals become":** Pfeiffer, *The Creative Explosion*, 114.

61 **Freud:** Brill, ed., *The Basic Writings of Sigmund Freud*.

61 **Frazer:** Frazer, *The Golden Bough*. See also Campbell, *Primitive Mythology*, 301–12.

61 **Eldredge:** Eldredge, *Dominion*, 91.

61 **Livingston:** Livingston, *One Cosmic Instant*, 137.

62 **climate:** Gamble, *The Paleolithic Settlement of Europe*, 89–94, 101–2.

62 **ice sheet:** Gamble, *The Paleolithic Settlement of Europe*, 90; Bogucki, *The Origins of Human Society*, 80–83; Fagan, *The Journey from Eden*, 178; Jonathan Adams, "Eurasia in the Last 150,000 Years," www.esd.ornl.gov/projects/qen/nercEURASIA.html;

GISP2 Home Page, gisp2.sr.unh.edu.69; ARCSS/GISP2 Ice Core, at gust.sr.unh.edu/ GISP2/DATA.

62 two refuge regions: Jochim, "Late Pleistocene Refugia in Europe"; Olga Soffer, "Upper Paleolithic Connubia, Refugia, and the Archeological Record from Eastern Europe," Soffer, ed., *The Pleistocene Old World*; Gamble, *The Paleolithic Settlement of Europe*, 213.

63 game shortages: Jochim, "Late Pleistocene Refugia in Europe," 321–25; Soffer, "Upper Paleolithic Connubia, Refugia, and the Archaeolgical Record from Eastern Europe," 343.

63 die-outs: Paul S. Martin, "Prehistoric Overkill: The Global Model," Martin and Klein, eds., *Quaternary Extinctions*, 385.

63 85 per cent: Pfeiffer, *The Creative Explosion*, 148, and see 58–59.

63 Spencer Wells: Wells, *The Journey of Man*, 132–34.

63 "trace their ancestry": Wells, *The Journey of Man*, 134.

64 Aurignacian toolkits: Klein, *The Human Career*, 481–84.

64 increasing number: Jochim, "Late Pleistocene Refugia in Europe"; Gamble, *The Paleolithic Settlement of Europe*, 223.

64 Bar-Yosef: Bar-Yosef, "Symbolic Expressions in Later Prehistory of the Levant," 176, 177.

65 burials: Klein, *The Human Career*, 550–53; Gamble, *The Paleolithic Settlement of Europe*, 186–89, 197; Bogucki, *The Origins of Human Society*, 97–98; B. Klima, "A Triple Burial from the Upper Paleolithic of Dolní Věstonice, Czechoslovakia," *Journal of Evolution* 16 (1987): 831–35; "Archeological Sites," hominids.com/donsmaps/indexsites.html.

65 Sungir: Randall White, Institute for Ice Age Studies (Insticeagestudies.com/library/ Ivory/ivory4); Olga Soffer, "Sungir: A Stone Age Burial Site," Goren Burenhult, ed., *The First Humans*, vol. 1 (New York: Harper Collins, 1993), 138–43; Tattersall, *Becoming Human*, 46; Klein, *The Human Career*, 551–53; Hominids.com/donsmaps/sungaea .html.

68 Grotte des Enfants, La Madeleine: Pfeiffer, *The Creative Explosion*, 67–68.

68 Mal'ta: Klein, *The Human Career*, 552.

69 Boehm: Christopher Boehm, *Hierarchy in the Forest: The Evolution of Egalitarian Behavior* (Cambridge: Harvard University Press, 1999), 4; also special issue of *Human Nature* 10 (1999): 205–52.

69 Pfeiffer: Pfeiffer, *The Creative Explosion*, 206–9.

Chapter Three: Intensification and Agriculture

71 last glacial maximum: Fagan, *The Journey from Eden*, 150–51; Klein, *The Human Career*, 58–61; Gamble, *Timewalkers*, 185–87, and *The Paleolithic Societies of Europe*, 282–87; Soffer, "The Middle to Upper Palaeolithic Transition on the Russian Plain," 721; Jonathan Adams, "Eurasia in the Last 150,000 Years."

71 ice sheets: Bogucki, *The Origins of Human Society*, 83.

72 refuge areas: Olga Soffer, "Upper Paleolithic Connubia, Refugia, and the Archaeo-logical Record from Eastern Europe," Soffer, ed., *The Pleistocene Old World*, 333–35; Michael Jochim, "Late Pleistocene Refugia in Europe," Soffer, ed., *The Pleistocene Old World*, 317–31; James A. Brown, "The Case for the Regional Perspective: A New World View," Soffer, ed., *The Pleistocene Old World*, 365–75.

72 occupancy, Cantabria and Perigord: Gamble, *The Paleolithic Settlement of Europe*, 223.

72 Russian scientists: N. K. Vereshchagin and G. F. Baryshnikov, "Quaternary Mam-malian Extinctions in Northern Eurasia," Martin and Klein, eds., *Quaternary Extinc-tions*, 508.

73 25–50: e.g. Fagan, *The Journey from Eden*, 186; Ezra Zubrow, "The Demographic Modelling of Neanderthal Extinction," Mellars and Stringer, eds., *The Human Revo-lution*, 214; Gamble, *The Paleolithic Settlement of Europe*, 50.

73 Soffer: Soffer, ed., *The Pleistocene Old World*, 333.

73 Jochim: Soffer, ed., *The Pleistocene Old World*, 318, 329, 326.

74 open-air sites, "climatic stress": Soffer, ed., *The Pleistocene Old World*, 340.

74 two paleoanthropologists: G. A. Clark and J. M. Lindly, "The Case of Continuity: Observations on the Biocultural Transition in Europe and Western Asia," Mellars and Stringer, eds., *The Human Revolution*, 643, 665.

74 "finest examples": Robert A. Guisepi, "Solutrean," ragz-international.com/stone_age.

74 spear points: Klein, *The Human Career*, 528–29; Lawrence Guy Straus, "Upper Paleolithic Hunting Tactics and Weapons in Western Europe," Peterkin, Bricker, and Mellars, eds., *Hunting and Animal Exploitation*, 88–90.

75 70 per cent: Straus, "Upper Paleolithic Hunting Tactics and Weapons in Western Europe," 88–90.

75 spear thrower: Straus, "Upper Paleolithic Hunting Tactics and Weapons in Western Europe," 89; Christopher A. Bergman, "The Development of the Bow in Western Europe: A Technological and Functional Perspective," Peterkin, Bricker, and Mellars, eds., *Hunting and Animal Exploitation*, 103; Klein, *The Human Career*, 540, 542; anno-tated Atlatl bibliography at web.grinnel.edu/anthropology/atlatl; Pierre Cattelain and Marie Perpere, "Tir expérimental de sagaies et de flèches emmanchées de pointes de la gravette," *Archeo-Situla* 17–20 (1993): 5–28; Heidi Knecht, ed., *Projectile Technology* (New York: Plenum, 1997).

75 modern replicas: annotated Atlatl bibliography at web.grinnel.edu/anthropology/atlatl.

75 oldest: Strauss, "Upper Paleolithic Hunting Tactics and Weapons in Western Europe"; Bergman, "The Development of the Bow in Western Europe."

75 bow and arrow: Bergman, "The Development of the Bow in Western Europe"; Klein, *The Human Career*, 541–42.

76 **"foreshafts":** Klein, *The Human Career*, 541.

76 **would "fit":** Bergman, "The Development of the Bow in Western Europe," 103.

76 **Cosquer:** Clottes and Courtin, *The Cave beneath the Sea*, 113, 141–45, dates at 170.

76 **Cougnac, Niaux:** Iain Davidson, "The Power of Pictures," Conkey et al., *Beyond Art*, 149.

76 **harpoons:** Klein, *The Human Career*, 538; Gail Larsen Peterkin, "Lithic and Organic Hunting Technology in the French Upper Palaeolithic," Peterkin, Bricker, and Mellars, eds., *Hunting and Animal Exploitation*, 59–62.

77 **boomerang:** P. Valde-Nowak et al., "Upper Palaeolithic Boomerang Made of a Mammoth Tusk in South Poland," *Nature* 329 (1 October 1987): 436–38; Paul G. Bahn, "Flight into Pre-history," *Nature* 373 (16 February 1995): 562; Per Michelson and Tasja W. Zhersok, "Bradshaw Rock Art" (Bradshawfoundation.com/bradshaws).

77 **Pavlov:** James M. Adovasio et al., "Upper Paleolithic Fibre Technology: Interlaced Woven Finds from Pavlo I, Czech Republic, c. 26,000 Years Ago," *Antiquity* 70, no. 269 (September 1996): 526–34; Olga Soffer and James M. Adovasio, at iabrno.cz/pv/soffadov.pdf.

77 **fencing:** Straus, "Upper Paleolithic Hunting Tactics and Weapons in Western Europe," Peterkin, Bricker, and Mellars, eds., *Hunting and Animal Exploitation*, 90.

78 **"regular, planned":** Lawrence Guy Straus, "Hunting in Late Upper Paleolithic Western Europe," Mathew H. Nitecki and Doris V. Nitecki, eds., *The Evolution of Human Hunting* (New York: Plenum, 1987) 147–6.

79 **cliff drives:** Klein, *The Human Career*, 463; Pfeiffer, *The Creative Explosion*, 59–61.

79 **Solutré:** Straus, "Hunting in Late Upper Paleolithic Western Europe," 84; Pfeiffer, *The Creative Explosion*, 60.

79 **"Les Trappes":** Straus, "Hunting in Late Upper Paleolithic Western Europe," 86.

79 **"masses of bones":** Straus, "Hunting in Late Upper Paleolithic Western Europe," 86.

79 **surrounds, Les Eylies, La Riera:** Straus, "Hunting in Late Upper Paleolithic Western Europe," 84–88.

79 **bison to deer:** Gamble, *The Paleolithic Settlement of Europe*, 108; Pfeiffer, *The Creative Explosion*, 61.

80 **France, large sites:** Randall White, "Glimpses of Long-Term Shifts in Late Paleolithic Land Use in the Perigord," Soffer, ed., *The Pleistocene Old World*, 268.

80 **Pastou Cliff:** Straus, "Hunting in Late Upper Paleolithic Western Europe," 86; Pfeiffer, *The Creative Explosion*, 197–98.

80 **creatures of habit:** Fagan, *The Journey from Eden*, 83.

80 **storage:** Gamble, "Man the Shoveler: Alternative Models for Middle Pleistocene Colonization and Occupation in Northern Latitudes," Soffer, ed., *The Pleistocene Old World*, 87–89; also Gamble, *The Paleolithic Settlement of Europe*, 107–9, 390–91.

81 **Woodburn:** Gowdy, ed., *Limited Wants, Unlimited Means*, 88–89.

81 Russian plain: Shreeve, *The Neandertal Enigma*, 317.

81 Pfeiffer: Pfeiffer, *The Creative Explosion*, 61.

81 "triple burial": Gamble, *The Paleolithic Societies of Europe*, 409–11; Shreeve, *The Neandertal Enigma*, 264–66.

82 Grimaldi: Klein, *The Human Career*, 555.

82 Wadi Kubbaniya: Fred Wendorf and Romuald Schild, *The Wadi Kubbaniya Skeleton* (Dallas: Southern Methodist University Press, 1986); Keeley, *War before Civilization*, 37; Klein, *The Human Career*, 555–56.

82 "got him": EBSCO Full Display (ehostvgw.7.epnet.com/fulltewxt.asp?resultSetId=Roo 000003&hitNum=2).

82–83 speared figures, "Killed Man": Clottes and Courtin, *The Cave beneath the Sea*, 155–61; Pfeiffer, *The Creative Explosion*, 105, 124.

83 six sites: Keith F. Otterbein, "The Origins of War," *Critical Review* 11, no. 2 (spring 1997): 255.

84 Wilson: Edward O. Wilson, *On Human Nature* (Cambridge: Harvard University Press, 1978), 82.

84 "those societies": Otterbein, "The Origins of War," 255, 257.

84 Jebel Sahaba: Fred Wendorf, *The Prehistory of Nubia* (Dallas: Southern Methodist University Press, 1968); Klein, *The Human Career*, 556; Keeley, *War before Civilization*, 37, 202; Gamble, *Timewalkers*, 190; R. Brian Ferguson, "The Causes and Origins of 'Primitive Warfare': On Evolved Motivations for War," *Anthropological Quarterly* 73, no. 3 (2000): 159–64; Arther Ferrill, "Neolithic War," eserver.org/history/neolithic-war.txt. (Dates vary in different accounts; Wendorf suggests 12,000 years ago, but it could be younger.)

84 extinctions: see in general Martin and Klein, eds., *Quaternary Extinctions*; Extinction Conference, American Museum of Natural History, April 1997, Amnh.org/science/biodiversity/extinction/Day 1; Anthony Barnosky et al., "Assessing the Causes of Late Pleistocene Extinctions on the Continents," *Science*, 1 October 2004, 70–75; Todd Surovell et al., "Global Archaeological Evidence for Proboscidean Overkill," *Proceedings of National Academy of Sciences* 102, no. 17 (26 April 2005): 6231–36.

85 60 per cent: Paul S. Martin, "Prehistoric Overkill: The Global Model," Martin and Klein, eds., *Quaternary Extinctions*, 384–86, plus cave bear, lion, and primitive bison.

85 mammoth: Martin, "Prehistoric Overkill," 387; Mithin, "Simulating Mammoth Hunting and Extinction: Implications for the Late Pleistocene of the Central Russian Plain," Peterkin, Bricker, and Mellars, eds., *Hunting and Animal Exploitation*; Diamond, *The Third Chimpanzee*, 313–29, 341–48; Claudine Cohen, *The Fate of the Mammoth* (Chicago: University of Chicago Press, 1994); Sergey A. Zimov, "Pleistocene Park: Return of the Mammoth's Ecosystem," *Science*, 6 May 2005, 796–98.

85 Chauvet: Jean-Marie Chauvet et al., *Dawn of Art: The Chauvet Cave* (New York: Harry N. Abrams, 1996).

85 Cosquer: Clottes and Courtin, *The Cave beneath the Sea.*

85 Owen-Smith: Norman Owen-Smith, "The Interaction of Humans, Megaherbivores, and Habitats in the Late Pleistocene Extinction Event," Extinction Conference, American Museum of Natural History, April 1997.

85 gestation: Diamond, *The Third Chimpanzee,* 346–47.

86 "the required hunting": Extinction Conference, 176.

87 European extinctions: Martin, "Prehistoric Overkill," 384–91; Vereshchagin and Baryshnikov, "Quaternary Mammalian Extinctions in Northern Eurasia"; Itzikoff, *The Inevitable Domination by Man,* 8–13.

87 Australian extinctions: Martin, "Prehistoric Overkill," 376–80; Klein, *The Human Career,* 571–72; "Australia, New Zealand, and the Island Pacific: Severe Losses," Martin and Klein, eds., *Quaternary Extinctions,* esp. essays by Peter Murray, David R. Horton, Geoffrey Hope, and A. Peter Kershaw; Richard G. Roberts, "Progress towards Single-Grain Optical Dating of Fossil Mud-Wasp Nests and Associated Rock Art in Northern Australia," *Quaternary Science Reviews* 22, nos. 10–13 (May 2003): 1273–78; James M. Bowler et al., "New Ages for Human Occupation and Climatic Change at Lake Mungo, Australia," *Nature,* 20 February 2003; Gifford H. Miller et al., "Ecosystem Collapse in Pleistocene Australia and a Human Role in Megafaunal Extinction," *Science,* 8 July 2005, 287–90.

88 fire setting: Klein, *The Human Career,* 572.

88 some paleologists: e.g. David R. Horton, "Red Kangaroos: Last of the Australian Megafauna," Martin and Klein, eds., *Quaternary Extinctions.*

88 Martin: Martin and Klein, eds., *Quaternary Extinctions.*

88 "killer possum": Murray, "Extinctions Downunder: A Bestiary of Extinct Australian Late Pleistocene Monotremes and Marsupials," Martin and Klein, eds., *Quaternary Extinctions,* 603.

88 "the arrival": Martin, "Prehistoric Overkill," 378.

88 American extinctions: Martin, "Prehistoric Overkill," 358–76; S. David Webb, "Ten Million Years of Mammal Extinctions in North America," Martin and Klein, eds., *Quaternary Extinctions,* 189–210; John E. Guilday, "Pleistocene Extinction and Environmental Change: Case Study of the Appalachians," Martin and Klein, eds., *Quaternary Extinctions,* 250–58; James E. King and Jeffrey Saunders, "Environmental Insularity and the Extinction of the American Mastodon," Martin and Klein, eds., *Quaternary Extinctions,* 315–44; Jim I. Mead and David J. Meltzer, "North American Late Quaternary Extinctions and the Radiocarbon Record," Martin and Klein, eds., *Quaternary Extinctions,* 440–50; Klein, *The Human Career,* 564–65; *Science,* 8 June 2001; Richard A. Kerr, "Megafauna Died from Big Kill, Not Big Chill," *Science,* 9 May 2003, 885.

88 ice sheets: Klein, *The Human Career,* 560.

88 Alaskan sites: Klein, *The Human Career*, 560–61; Ted Goebel et al., "The Archae-ology of Ushki Lake, Kamchatka, and the Pleistocene People of the Americas," *Science* (25 July 2003): 501–5; Richard Stone, "Late Date for Siberian Site Challenges Bering Pathway," *Science*, 25 July 2003, 450–51.

89 Meadowcroft, Monte Verde: Klein, *The Human Career*, 563–64; Diamond, *The Third Chimpanzee*, 341; D. J. Meltzer, "Monte Verde and the Pleistocene People of the Ameri-cas," *Science* 276 (2 May 1997): 754.

89 Pikamachay: Martin, "Prehistoric Overkill," 375.

89 Taima-Taima: Ruth Gruhn and Alan L. Bryan, "The Record of Pleistocene Mega-faunal Extinctions at Taima-taima, Northern Venezuela," Martin and Klein, eds., *Quaternary Extinctions*, 128–37.

89 thirty-five genera: Gary Haynes, *The Early Settlement of North America* (Cambridge: Cambridge University Press, 2002), reviewed by Vance T. Holliday, "Where Have All the Mammoth Gone?," *Science*, 30 May 2003, 1373–74.

89 50 to 100 million: Ehrenreich, *Blood Rites*, 119.

89 80 per cent: Martin, "Prehistoric Overkill," 358, 370–76.

90 carnivore extinction theory: Elin Whitney-Smith, "The End of Eden," well.com/user/elin/edentxt.

90 Klein: Klein, *The Human Career*, 565.

90 extinction percentages: Martin, "Prehistoric Overkill," 358; Alden Peterson, "Human Cultural Agency in Extinction," *Wild Earth*, spring 2003, 10–13; A. J. Stuart, "The Role of Humans in Late Pleistocene Megafaunal Extinction, with Particular Reference to Northern Eurasia and North America," Extinction Conference, American Museum of Natural History, April 1997, 86.

90 Crosby: Alfred Crosby, *Throwing Fire* (Cambridge: Cambridge University Press, 2002), 52.

90 Klein: Klein and Edgar, *The Dawn of Human Culture*, 252; similar at Klein, *The Human Career*, 572.

90–91 island extinctions: Martin, "Prehistoric Overkill," 391–94; Richard Cassels, "Fau-nal Extinction and Prehistoric Man in New Zealand and the Pacific Islands," Martin and Klein, eds., *Quaternary Extinctions*, 741–67; Storrs. L. Olson and Helen F. James, "The Role of Polynesians in the Extinction of the Avifauna of the Hawaiian Islands," Martin and Klein, eds., *Quaternary Extinctions*, 768–84; Jared M. Diamond, "Historic Extinction: A Rosetta Stone for Understanding Prehistoric Extinctions," Martin and Klein, eds., *Quaternary Extinctions*, 824–66.

91 Wilson: Wilson, *The Future of Life*, 94, 102.

91 climate 16,000–11,000: Gamble, *The Paleolithic Settlement of Europe*, 213; Adams, "Eurasia in the Last 150,000 Years."

92 population pressure and resettlement: Cohen, *The Food Crisis in Prehistory*, esp. chapters 2, 3; Ehrlich, *Human Natures*, 403; Soffer, "Upper Paleolithic Connubia, Refugia, and the Archaeological Record from Eastern Europe," 344; Gamble, *The Paleolithic Settlement of Europe*, 205, 223.

92 Africa: Cohen, *The Food Crisis in Prehistory*, 102–4.

92 "a striking number": Cohen, *The Food Crisis in Prehistory*, 102.

92 Levant: Cohen, *The Food Crisis in Prehistory*, 137.

92 one estimate: Cohen, *The Food Crisis in Prehistory*, 138; Harris, *Cannibals and Kings*, 29, estimates 100,000 people at 8,000 years ago.

92 carbon dioxide: Rowan Sage, of the University of Toronto, in Robert C. Balling, "Origins of Agriculture," Greeningearthsociety.org/Articles/origins, and Richerson et al., "Origins of Agriculture," des.ucdavis.ed; faculty/Richerson/origins_ag_IV3.

93 sickle sheen: Cohen, *The Food Crisis in Prehistory*, 135, 105; Pfeiffer, *The Creative Explosion*, 198.

93 grindstones: Cohen, *The Food Crisis in Prehistory*, 101, 105, 135, 153, 218.

93 year's supply: Jack R. Harlan, cited in Cohen, *The Food Crisis in Prehistory*, 142.

94 "very unlikely": Cohen, *The Food Crisis in Prehistory*, 147.

94 Fertile Crescent: Ofer Bar-Yosef, "The Contributions of Southwest Asia to the Study of the Origin of Modern Humans," Nitecki and Nitecki, eds., *Origins of Anatomically Modern Humans*, 52–53; Ofer Bar-Yosef, "The Natufian Culture in the Levant, Threshold to the Origins of Agriculture," *Evolutionary Anthropology* 6, no. 5 (7 December 1998): 159–77; Diamond, *Guns, Germs, and Steel*, 126, 134–42; Jared Diamond, "Evolution, Consequences and Future of Plant and Animal Domestication," *Nature*, 8 August 2002, 700.

95 Natufian: Bar-Yosef, "The Contributions of Southwest Asia to the Study of the Origin of Modern Humans," 52–53; Bar-Yosef, "The Natufian Culture in the Levant"; Ofer Bar-Yosef, "Symbolic Expressions in Later Prehistory of the Levant: Why Are They So Few?," Conkey et al., eds., *Beyond Art*, 168–71.

95 stone and clay buildings: Cohen, *The Food Crisis in Prehistory*, 135.

95 extinctions: Eitan Tchernov, "Faunal Turnover and Extinction Rate," Martin and Klein, eds., *Quaternary Extinctions*, esp. 540; Diamond, *Guns, Germs, and Steel*, 142.

95 sickles: Alexander Marshack, "Paleolithic Image Making and Symboling in Europe and the Middle East: A Comparative Review," Conkey et al., *Beyond Art*, 75, 78.

95 storage, mortars, slabs, bowls: Cohen, *The Food Crisis in Prehistory*, 134–35; Harris, *Cannibals and Kings*, 26; Bar-Yosef, "The Natufian Culture in the Levant," 162, 164–65.

95 "Natufian communities": Bar-Yosef, "The Natufian Culture in the Levant," 167.

95 climate change ("Younger Dryas"): David W. Lea et al., "Synchroneity of Tropical and High-Latitude Atlantic Temperatures over the Last Glacial Termination," *Science*, 5 September 2003, 1361; R. B. Alley et al., "Abrupt Climate Change," *Science*,

28 March 2003, 2005–10; ARCSS/GISP2 Ice Core, 91; Harvey Weiss, Garth Bawden, and Richard Reycraft, *Confronting Natural Disaster* (Albuquerque: University of New Mexico Press, 2000).

96 fifty sites, scenario, "very probably": John Noble Wilford, "Grains Resemble Their Wild Cousins, with a Few Crucial Differences: New Clues Show Where People Made the Leap to Agriculture," *New York Times*, 18 November 1997, § F, 1–2; Manfred Heun et al., "Site of Einkorn Wheat Domestication Identified by DNA Fingerprinting," *Science*, 14 November 1997, 1312–14. See also David Rindos, *The Origins of Agriculture* (New York: Academic, 1984); John Pfeiffer, "The Rise of Farming in the Near East," *The Emergence of Society* (New York: McGraw-Hill, 1977), 127–48; Clive Ponting, "The First Great Transition," *A Green History of the World* (New York: St. Martin's, 1992), 37–67.

96 other sites: Martin K. Jones et al., "Wheat Domestication," *Science*, 16 January 1998, 302–4; Wilford, "Grains Resemble Their Wild Cousins."

96 other areas: Diamond, *Guns, Germs, and Steel*, chapters 5–8, esp. 98–103.

96 China: Richerson et al., *The Origin and Evolution of Cultures* (New York: Oxford University Press, 2005).

96 Andes: Dolores R. Piperno and Karen E. Stothert, "Phytolith Evidence for Early Holocene Cucurbita Domestication in Southwest Ecuador," *Science*, 14 February 2003, 1054–57; Cohen, *The Food Crisis in Prehistory*, 228.

97 Mexico: Bruce Smith, USANews.com, 18 June 2001 (crystalinks.com/agriculture history.html).

97 New Guinea: Katharina Neumann, "New Guinea: A Cradle of Agriculture," *Science*, 11 July 2003, 180–81.

97 Egypt, Africa: Cohen, *The Food Crisis in Prehistory*, 108.

97 "We began": Hillman, crystalinks.com/agriculturehistory.html.

97 "We removed": Eldredge, *Dominion*, 93.

98 domestication of animals: Diamond, *Guns, Germs, and Steel*, 169–74, and "Evolution, Consequences and Future of Plant and Animal Domestication."

98 domestication criteria: Diamond, *Guns, Germs, and Steel*, 168–74.

98 Abu Hureyra: David E. MacHugh and Danile G. Bradley, "Livestock Genetic Origins: Goats Buck the Trend," *Proceedings of the National Academy of Sciences* 98, no. 10 (8 May 2001): 5382–84; Andrew M. T. Moore et al., *Village on the Euphrates: The Excavation of Abu Hureyra* (Oxford: Oxford University Press, 1999); "Early Food Production in the Old World," *Encyclopedia of World History* (Boston: Houghton Mifflin, 2001); Wilford, "Grains Resemble Their Wild Cousins."

98 Ganj Dareh, Ali Kosh: Melinda A. Zeder and Brian Hesse, "The Initial Domestication of Goats (Capra Hircus) in the Zargos Moutains 10,000 Years Ago," *Science*, 24 March 2000, 2254–57.

98 Umm Qseir: Melinda A. Zeder, mc.maricopa.edu/dept/d10/asb/learning/lifeways.

99 "the worst mistake": Jared Diamond, "The Worst Mistake in Human History," *Discover*, May 1987, 64–66. See also Zerzan, *Elements of Refusal*, 73–87; Ehrlich, *Human Natures*, 236–40; Lawrence J. Angel, Mark N. Cohen, and George J. Armelagos, eds., *Paleopathology at the Origins of Agriculture* (New York: Academic, 1984); Bogucki, *The Origins of Human Society*, 195–202.

99 population numbers: Diamond, "The Worst Mistake in Human History."

100 cities: Pfeiffer, "The Coming of Cities and Kings," *The Emergence of Society*, 149–70; Lewis Mumford, *The City in History* (New York: Harcourt Brace, 1961), 41–46, 55–57; Bogucki, *The Origins of Human Society*, 230–36; Fellipe Fernandez-Armesto, *Civilizations* (New York: Free Press, 2001), 180–83; James Mellaart, *Catal Huyuk: A Neolithic Town in Anatolia* (New York: McGraw-Hill, 1967); Kathleen Kenyon, *Jericho* (New York: Praeger, 1957); William Eichman, "Catal Huyuk: The Temple City of Prehistoric Anatolia," *Gnosis* 15 (spring 1990): 52–59; bibliography at catal.arch.cam.ac.uk/catal/catal.

100 private property: Stringer and McKie, *African Exodus*, 234.

100 diseases: Diamond, *Guns, Germs, and Steel*, 205–7; Diamond, "Evolution, Consequences and Future of Plant and Animal Domestication"; Nicholas Wade, "Gene Study Dates Malaria to the Advent of Farming," *New York Times*, 22 June 2001, § A, 18; T. A. Cockburn, "Infectious Diseases in Ancient Populations," *Current Anthropology* 12, no. 1 (1971): 45–62; P. W. A. Edward, *Evolution of Infectious Disease* (Oxford: Oxford University Press, 1994); Cohen, *The Food Crisis in Prehistory*, 29.

101 life spans: Klein, *The Human Career*, 553–54; Diamond, "The Worst Mistake in Human History."

101 "humans": Stringer and McKie, *African Exodus*, 233.

101 body height: Stringer and McKie, *African Exodus*, 231; Diamond, "The Worst Mistake in Human History"; he gives the heights of Sapiens hunters as 5′ 9″ for men, 5′ 5″ for women.

101 average height today: Stringer and McKie, *African Exodus*.

101 brain size: Stringer and McKie, *African Exodus*; Diamond, "Evolution, Consequences and Future of Plant and Animal Domestication."

101 "humans, through": William F. Allman, *Evolving Brains* (New York: W. H. Freeman, 2000).

101–102 inferior food, one study: Diamond, "The Worst Mistake in Human History."

102 division of labor, hierarchy: Zerzan, *Elements of Refusal*; Pfeiffer, *The Emergence of Society*; James DeMeo, *Saharasia: The 4000 BCE Origins of Child Abuse, Sex Repression, Warfare, and Social Violence in the Deserts of the Old World* (Ashland, Ore.: Natural Energy, 1998); Diamond, *The Third Chimpanzee*, 187–89; John Noble Wilford, "Civilization's Cradle Grows Larger," *New York Times*, 28 May 2002, § F, 2; Jared Diamond and Peter Bellwood, "Farmers and Their Languages: The First Expansions," *Science*, 25 April 2003, 597–604.

103 "Forced to choose": Diamond, "The Worst Mistake in Human History."

103 spread of agriculture: Diamond and Bellwood, "Farmers and Their Languages"; Diamond, *Guns, Germs, and Steel*, 178–91; Diamond, *The Third Chimpanzee*, 189–90.

103 failure of agriculture: Clive Ponting, "Destruction and Survival," *A Green History of the World*, 68–87; Marc Reisner, *Cadillac Desert* (New York: Viking, 1986), 474–80, 486–87.

Chapter Four: The Erectus Alternative

105 Homo erectus: Klein, *The Human Career*, 267–75, 280–95.

105 Turkana Boy: Walker and Shipman, *The Wisdom of the Bones*.

106 younger age: Leakey, *The Origin of Humankind*, 51.

106 walking, running: Leakey, *The Origin of Humankind*, 56.

107 one study: Roger Lewin, *Principles of Human Evolution* (Malden, Mass.: Blackwell Science, 1998), 452.

107 Erectus and early Sapiens brain size: Klein, *The Human Career*, 276, table and note.

107 30 per cent: compared to Homo habilis—Klein, *The Human Career*, 218.

107 brain tissue: William Noble and Iain Davidson, "The Evolutionary Emergence of Modern Human Behavior," *Man* 26, no. 2 (June 1991): 223–53.

107 Robert Martin: Leakey, *The Origin of Humankind*; Stringer and McKie, *African Exodus*, 55.

108 fire: Naama Goren-Inbar et al., "Evidence of Hominid Control of Fire at Gersher Benot Ya'agov, Israel," *Science*, 30 April 2004, 725–27 (Levant); Klein, *The Human Career*, 351 (China).

108 FxJj20: Ralph Rowlett et al., discoveringarchaelogy.com/0599toc/5feature3-fire .shtml.

109 Wrangham: Richard Wrangham et al., "The Raw and the Stolen: Cooking and the Ecology of Human Origins," *Current Anthropology* 40, no. 5 (December 1999): 567–94.

109 "the beginnings": Klein, *The Human Career*, 293.

110 "phenomenal": Klein, *The Human Career*, 280.

110 "Even as": Walker and Shipman, *The Wisdom of the Bones*, 163.

110 stature: Klein, *The Human Career*, 191.

110 tools: Klein, *The Human Career*, 327–28, 333–44, 346.

110 Wynn: Wynn, *The Evolution of Spatial Competence*.

111 "a frame": Wynn, *The Evolution of Spatial Competence*, 57.

111 "sophisticated notions": Wynn, *The Evolution of Spatial Competence*, 67.

111 "hominids possessed": Wynn, *The Evolution of Spatial Competence*, 92.

112 Ceprano skull: Klein, *The Human Career*, 323.

112 **sexual selection:** Miller, *The Mating Mind*, esp. 35–48 on Darwin.

112 **dispersal:** Klein, *The Human Career*, 267–75, 314–27.

112 **50 sites:** Klein, *The Human Career*, 265, 334.

112 **Japan:** archeology.org/online/news/home, 24 March 2000.

112 **population:** Shahin Rouhani, "Molecular Genetics and the Pattern of Human Evolution: Plausible and Implausible Models," Mellars and Stringer, eds., *The Human Revolution*, 49.

113 **Site 50:** Leakey, *The Origin of Humankind*; Wynn, *The Evolution of Spatial Competence*; Stringer and McKie, *African Exodus*, 68–71; Glynn Isaac, "Food-Sharing Behavior of Protohuman Hominids," *Scientific American* 238, no. 4 (April 1978): 90.

114 **neural cells:** Walker and Shipman, *The Wisdom of the Bones*, 211–14; Ann MacLarnen, Alan Walker, and Richard Leakey, eds., *The Nariokotome Homo Erectus Skeleton* (Cambridge: Harvard University Press, 1993); Klein, *The Human Career*, 348.

114 **group size, limits:** Gamble, *The Paleolithic Settlement of Europe*, 50; Gamble, *The Paleolithic Societies of Europe*, 57; Pfeiffer, *The Creative Explosion*, 192–94.

115 **convinced him:** Walker and Shipman, *The Wisdom of the Bones*, 185.

115 **KNM-ER 1808:** Walker and Shipman, *The Wisdom of the Bones*, 128–36, 157.

115 **"*Someone else*":** Walker and Shipman, *The Wisdom of the Bones*, 134.

115 **"poignant testimony":** Walker and Shipman, *The Wisdom of the Bones*, 134.

115 **carnivore liver:** Walker and Shipman, *The Wisdom of the Bones*, 136–37.

116 **skeleton numbers:** Klein, *The Human Career*, 265, 314–27; "Prominent Hominid Fossils," talkorigins.org.

116 **"immediate-return" societies:** see esp. Gowdy, ed., *Limited Wants, Unlimited Means*, 87–110, xxix–xxxi.

116 **"a remarkable degree":** Gowdy, ed., *Limited Wants, Unlimited Means*, 90.

117 **"look on any":** Kevin Duffy, *Children of the Forest: Africa's Mbuti Pygmies* (Prospect Heights, Ill.: Waveland, 1984), quoted by John Zerzan, "Future Primitive," in Gowdy, ed., *Limited Wants, Unlimited Means*, 268.

117 **acceptance of death, "When you're dead":** Berman, *Wandering God*, 76–77.

118 **Turnbull:** quoted in Berman, *Wandering God*, 77.

119 **Klein:** Klein, *The Human Career*, 246.

119 **"where scavenging," "substantial amounts":** Klein, *The Human Career*, 247, based on Ralph Blumenschine, *Early Hominid Scavenging Opportunities: Implications of Carcass Availability in the Serengeti and Ngorongoro Ecosystems* (Oxford: B.A.R., 1986).

119 **meat as a third of diet:** Richard B. Lee, "What Hunters Do for a Living, or, How to Make Out on Scarce Resources," Lee and Irven DeVore, eds., *Man the Hunter* (Chicago: Aldine, 1968); Gowdy, ed., *Limited Wants, Unlimited Means*, 59–60.

120 **saiga antelope:** Huntingfortomorrow.com fact sheet says that an antelope has 43 calories per ounce; Elaine Anderson, "Who's Who in the Pleistocene: A Mammalian Besti-

ary," Martin and Klein, eds., *Quaternary Extinctions*, 81, gives an average weight of 52 kilograms, or 115 pounds (1840 ounces) for saiga antelopes, which works out to 79,120 calories.

120 Barfield: Barfield, *Saving the Appearances*, 123, 124, 186; also see Stephen L. Talbott, *The Future Does Not Compute* (Sebastopol, Calif.: O'Reilly and Associates, 1995), appendix A.

120 Shreeve: Shreeve, *The Neandertal Enigma*, 340–41.

121 van der Post: Laurens van der Post, *The Lost World of the Kalahari* (New York: Vintage, 1958), quoted in Zerzan, *Running on Emptiness*, 14, and Zerzan, "Whose Future," *Species Traitor*, winter 2001, 20.

122 Shepard: Gowdy, ed., *Limited Wants, Unlimited Means*, 291–92.

123 Jung: *New Dimensions Annual Journal*, 2002, 19, in a conversation with Laurens van der Post.

123 Shepard: Gowdy, ed., *Limited Wants, Unlimited Means*, 314.

124 Havel: Vaclav Havel, "Address by His Excellency Vaclav Havel, President of the Czechoslovakian Socialist Republic," *Congressional Record*, 21 February 1990.

124 acceptance of death: Paul Shepard, "Post-Historic Primitivism," Gowdy, ed., *Limited Wants, Unlimited Means*, 314–15.

125 scale: Gamble, *The Paleolithic Settlement of Europe*, 50–51; Kirkpatrick Sale, *Human Scale* (New York: Coward McCann, 1980; Perigee, 1982), 182–83; John Pfeiffer, *The Emergence of Man* (New York: Harper, 1972), 376–77.

126 population density: Gamble, *The Paleolithic Societies of Europe*, 41.

126 Blumenfeld, Pfeiffer: Sale, *Human Scale*.

127 Bacon: quoted in Derek Jensen, *A Language Older than Words* (New York: Context, 2000), 20.

128 Jameson: Fredric Jameson, *Postmodernism, or, The Cultural Logic of Late Capitalism* (Durham: Duke University Press, 1991), ix, 34.

128 Wilson: Stephen R. Keller and Edward O. Wilson, eds., *The Biophilia Hypothesis* (Washington: Island/Shearwater, 1995), 31–32.

129 Loewald: quoted in Berman, *Wandering God*, 243.

129 Powys: John C. Powys, *The Art of Happiness* (New York: Simon and Schuster, 1935; London: Village, 1974), quoted in *Resurgence* 221 (November–December 2003).

129 LaConte: Ellen LaConte, "The Next Things after Surrender," *Trumpeter* (British Columbia) 19, no. 1 (2003): 97–108; rainforestinfo.org.au/deep-eco/ellen_laconte.htm.

130 Berry: Berry's summary of the "historical mission of our time," including "at the species level," *Ecozoic Reader*.

130 Berry: Berry, *The Dream of the Earth*, 202.

130 Deep Ecology: Bill Devall and George Sessions, *Deep Ecology* (Layton, Utah: Peregrine Smith, 1985), 70.

131 Leopold: quoted in *Wild Earth*, fall 1999.

131 "In the beginning": Akwesasne Notes, eds., *A Basic Call to Consciousness* (Rooseveltown, N.Y.: Mohawk Nation, 1978); original address to the United Nations, Geneva, 9 September 1977, on line at ewetribe.com/NAculture.

132 Morton: James Parks Morton, "The Organizing Principle of Human Consciousness," *Annals of Earth* 16, no. 3 (1998): 8, also in Frederick Franck et al., *What Does It Mean to Be Human?* (Nyack, N.Y.: Circumstantial Productions, 1998), 149–50.

133 New Cosmology: Thomas Berry and Brian Swimme, *The Universe Story* (New York: Harper Collins, 1994).

133 Lovelock: James Lovelock, *Gaia* (New York: Oxford, 1979).

133 McLaughlin: Andrew McLaughlin, *Regarding Nature: Industrialism and Deep Ecology* (Albany: SUNY Press, 1993), 14; see also Devall and Sessions, *Deep Ecology*, and Sessions, ed., *Deep Ecology for the Twenty-first Century* (Boston: Shambala, 1995).

133 bioregionalism: Kirkpatrick Sale, *Dwellers in the Land: The Bioregional Vision* (San Francisco: Sierra Club, 1985), quote from 173; see also Berry, *The Dream of the Earth*, chapters 12, 13, and Peter Berg, *Figures of Regulation* (San Francisco: Planet Drum Foundation, 1982).

133 animal rights: Peter Singer, "Animal Liberation at 30," *New York Review of Books*, 15 May 2003.

133 stewardship: Matthew Scully, *Dominion: The Power of Man, the Suffering of Animals, and the Call to Mercy* (New York: St. Martin's, 2002).

134 primitivism: Zerzan, *Elements of Refusal*; Zerzan, *Future Primitive*; Zerzan, "Future Primitive," Gowdy, ed., *Limited Wants, Unlimited Means*, 255–80; Paul Shepard, "A Post-Historic Primitivism," Gowdy, ed., *Limited Wants, Unlimited Means*, 281–326; Kim Stanley Robinson, ed., *Future Primitive: The New Ecotopias* (New York: TOR, 1994); Stanley Diamond, *In Search of the Primitive: A Critique of Civilization* (New Brunswick, N.J.: Transaction, 1974); Sahlins, *Stone Age Economics*; Primitivism.com. (with articles by Pierre Clastres, Stanley Diamond, John Filiss, Chellis Glendinning, Richard Heinberg, Lewis Mumford, Fredy Perlman, Paul Shepard, and others).

134 "Why would one": Zerzan, *Elements of Refusal*, 67.

134 "Number is the most": Zerzan, *Elements of Refusal*, 47.

135 "Genetic engineering": *Anarchy* (Columbia, Mo.), fall–winter 2003–4, 45.

135 "As I make it": Daniel Quinn, *Ishmael* (New York: Bantam, 1992), 126–27.

135 "I no longer think": Quinn, *Ishmael*, 130, and *Beyond Civilization: Humanity's Next Great Adventure* (New York: Harmony, 1999).

135 "Man the Hunter": Lee and Devore, eds., *Man the Hunter*, 99; Gowdy, ed., *Limited Wants, Unlimited Means*.

135 *Cambridge Encyclopedia*: Richard B. Lee and Richard Daly, eds., *The Cambridge Encyclopedia of Hunters and Gatherers* (Cambridge: Cambridge University Press, 1999).

136 Sahlins: Sahlins, *Stone Age Economics*; Gowdy, ed., *Limited Wants, Unlimited Means*, 32, 35.

136 "Mounting evidence": *Anarchy* (Columbia, Mo.), fall–winter 2003–4, 45.

136 Teaching Drum: *Earth First! Journal*, March–April 2002.

137 Society: Primitive.org. See also John and Geri McPherson, *"Naked into the Wilderness": Primitive Living and Survival Skills* (Randolph, Kan.: Prairie Wolf, 1993).

137 "I'm doing it": *Hudson Valley*, April 2000.

BIBLIOGRAPHY

Abram, David. *The Spell of the Sensuous*. New York: Pantheon, 1996.

Allman, William F. *The Stone Age Present*. New York: Simon and Schuster, 1994.

Barfield, Owen. *Saving the Appearances*. London: Faber and Faber, 1957.

Berman, Morris. *Coming to Our Senses*. New York: Simon and Schuster, 1989.

———. *Wandering God*. Albany: State University of New York Press, 2000.

Berry, Thomas. *The Dream of the Earth*. San Francisco: Sierra Club Books, 1988.

Binford, Lewis R. *In Pursuit of the Past: Decoding the Archaeological Record*. New York: Thames and Hudson, 1983.

Bogucki, Peter. *The Origins of Human Society*. Malden, Mass.: Blackwell, 1999.

Cartmill, Matt. *A View to a Death in the Morning*. Cambridge: Harvard University Press, 1993.

Cavalli-Sforza, Luigi Luca, et al. *The Great Human Diasporas*. New York: Addison-Wesley, 1995.

Clottes, Jean, and Jean Courtin. *The Cave beneath the Sea*. New York: Harry N. Abrams, 1996.

Cohen, Mark Nathan. *The Food Crisis in Prehistory*. New Haven: Yale University Press, 1977.

Conkey, Margaret W., et al., eds. *Beyond Art: Pleistocene Image and Symbol*. San Francisco: California Academy of Sciences, 1997.

Deacon, H. J., and Janette Deacon. *Human Beginnings in South Africa*. Walnut Creek, Calif.: Altamira, 1999.

Deloria, Vine, Jr. *Red Earth, White Lies*. Golden, Colo.: Fulcrum, 1997.

Diamond, Jared. *The Third Chimpanzee*. New York: Harper Collins, 1992.

———. *Guns, Germs, and Steel*. New York: W. W. Norton, 1997.

Dunbar, Robin. *Grooming, Gossip, and the Evolution of Language*. New York: Faber and Faber, 1996.

Ehrenreich, Barbara. *Blood Rites*. New York: Metropolitan, 1997.

Ehrlich, Paul R. *Human Natures*. Washington: Island, 2000.

Eldredge, Niles. *Dominion*. New York: Henry Holt, 1995.

Fagan, Brian M. *The Journey from Eden*. New York: Thames and Hudson, 1990.

Fisher, Helen E. *The Sex Contract*. New York: William Morrow, 1982.

Fox, Richard Gabriel. *The Neanderthal Problem and the Evolution of Human Behavior*. Chicago: University of Chicago Press, 1998.

———, ed. "The Neanderthal Problem and the Evolution of Human Behavior," *Current Anthropology* 39 (1998) [suppl.].

Frazer, James George. *The Golden Bough*. London: Macmillan, 1890.

Gamble, Clive. *The Paleolithic Settlement of Europe*. Cambridge: Cambridge University Press, 1986–96.

———. *Timewalkers: The Prehistory of Global Civilization*. Cambridge: Harvard University Press, 1993.

———. *The Paleolithic Societies of Europe*. Cambridge: Cambridge University Press, 1999–2000.

Goodman, Felicitas D. *Ecstasy, Ritual, and Alternate Reality*. Bloomington: Indiana University Press, 1988.

Gowdy, John, ed. *Limited Wants, Unlimited Means: A Reader on Hunter-Gatherer Economics and the Environment*. Washington: Island, 1998.

Hadingham, Evan. *Secrets of the Ice Age*. New York: Walker, 1979.

Harris, Marvin. *Cannibals and Kings*. New York: Random House, 1977.

Itzkoff, Seymour W. *The Inevitable Domination by Man*. Ashfield, Mass.: Paideia, 2000.

Keeley, Lawrence H. *War before Civilization*. Oxford: Oxford University Press, 1996.

Kelly, Robert L. *The Foraging Spectrum: Diversity in Hunterer-Gatherer Lifeways*. Washington: Smithsonian Institution Press, 1995.

Kingdon, Jonathan. *Self-Made Man*. New York: John Wiley and Sons, 1993.

Klein, Richard G. *The Human Career: Human Biological and Cultural Origins*. Chicago: University of Chicago Press, 1989; 2d ed., 1999.

Klein, Richard G., with Blake Edgar. *The Dawn of Human Culture*. New York: John Wiley and Sons, 2002.

Knecht, Heidi, Anne Pike-Tay, and Randall White, eds. *Before Lascaux: The Complex Record of the Early Upper Paleolithic*. Boca Raton: CRC, 1993.

Leakey, Richard. *The Origin of Humankind*. New York: Basic, 1994.

Leakey, Richard, and Roger Lewin. *Origins*. New York: E. P. Dutton, 1977.

Livingston, John A. *One Cosmic Instant*. Boston: Houghton Mifflin, 1973.

Martin, Paul S., and Richard G. Klein, eds. *Quaternary Extinctions*. Tucson: University of Arizona Press, 1984.

McBrearty, Sally, and Alison S. Brooks. "The Revolution That Wasn't," *Journal of Human Evolution* 39, no. 5 (November 2000): 453–563.

Mellars, Paul, ed. *The Emergence of Modern Humans*. Ithaca: Cornell University Press, 1990.

Mellars, Paul, and Chris Stringer, eds. *The Human Revolution*. Princeton: Princeton University Press, 1989.

Miller, Geoffrey. *The Mating Mind*. New York: Doubleday, 2000.

Morris, Desmond. *The Naked Ape*. New York: McGraw-Hill, 1967.

Morrison, Reg. *The Spirit in the Gene*. Ithaca: Cornell University Press, 1999.

Nitecki, Matthew H., and Doris V. Nitecki, eds. *Origins of Anatomically Modern Humans*. New York: Plenum, 1994.

Peterkin, Gail Larsen, Harvey M. Bricker, and Paul Mellars, eds. *Hunting and Animal Exploitation in the Later Palaeolithic and Mesolithic of Eurasia*. Washington: American Anthropological Association, 1993.

Pfeiffer, John. *The Creative Explosion*. New York: Harper and Row, 1982.

Rees, Martin. *Our Final Hour: A Scientist's Warning: How Terror, Error, and Environmental Disaster Threaten Humankind's Future in This Century on Earth and Beyond*. New York: Basic, 2003.

Rightmire, G. Philip, *The Evolution of Homo Erectus*. Cambridge: Cambridge University Press, 1990.

Rudgely, Richard. *Lost Civilisation of the Stone Age*. London: Century, 1998.

Sahlins, Marshall. *Stone Age Economics*. Chicago: Aldine-Atherton, 1972.

Santa Luca, A. P. *The Ngandong Fossil Hominids*. New Haven: Yale University, Department of Anthropology, 1980.

Shepard, Paul. *Nature and Madness*. San Francisco: Sierra Club Books, 1982.

Shreeve, James. *The Neandertal Enigma*. New York: William Morrow, 1995.

Soffer, Olga, ed. *The Pleistocene Old World*. New York: Plenum, 1987.

Stanley, Steven M. *Children of the Ice Age: How a Global Catastrophe Allowed Humans to Evolve*. New York: Harmony, 1996.

Stringer, Christopher, and Clive Gamble. *In Search of the Neanderthals*. New York: Thames and Hudson, 1993.

Stringer, Christopher, and Robin McKie. *African Exodus*. New York: Henry Holt, 1996.

Swisher, Carl C., Garniss H. Curtis, and Roger Lewin. *Java Man*. New York: Scribner, 2000.

Tattersall, Ian. *The Fossil Trail*. Oxford: Oxford University Press, 1995.

———. *Becoming Human*. New York: Harcourt, 1998.

———. *The Last Neanderthal*. Boulder: Westview, 1999.

———. *The Monkey in the Mirror*. New York: Harcourt, 2002.

Tattersall, Ian, and Jeffrey Schwartz. *Extinct Humans*. Boulder: Westview, 2001.

Taylor, Timothy. *The Prehistory of Sex*. New York: Bantam, 1996.

Turnbull, Colin. *The Forest People*. New York: Simon and Schuster, 1961.

Walker, Alan, and Pat Shipman. *The Wisdom of the Bones: In Search of Human Origins*. New York: Alfred A. Knopf, 1996.

Wells, Spencer. *The Journey of Man*. Princeton: Princeton University Press, 2002.

Wilson, E. O. *Consilience*. New York: Alfred A. Knopf, 1998.

———. *The Future of Life*. New York: Alfred A. Knopf, 2002.

Wynn, Thomas. *The Evolution of Spatial Competence*. Urbana: University of Illinois Press, 1989.

Zerzan, John. *Future Primitive*. New York: Autonomedia, 1994.

———. *Against Civilization*. Eugene, Ore.: Uncivilized Books, 1999.

———. *Elements of Refusal*. Seattle: Left Bank, 1988; 2d ed. Columbus, Mo.: C.A.L., 1999.

———. *Running on Emptiness*. Los Angeles: Feral House, 2002.

ACKNOWLEDGMENTS

For many kinds of assistance and encouragement, for which I am deeply grateful, I would like to thank the following: Ian Baldwin, Ed Barber, Ivy Bell, Adam Bellow, Steve Cohn, Warren Egerter, Clive Gamble, Chellis Glendinning, John Horgan, Sally McBrearty, Andrew McLaughlin, Valerie Millholland and her colleagues, John Paulits, Ray Reece, Norman Rush, Roger Sale, Gilbert Tostvin, Amanda Urban, and Hugh van Dusen.

I owe a special debt of gratitude for the support and companionship of my cherished partner, Shirley Branchini.

INDEX

tralia, 35–36, 87–88; in Europe, 7, 63, 79–80, 84–87; in Levant, 46

Fire, 21, 35–36, 60, 61, 88, 89, 108–9; farming with, 24–25
Fishing, 13, 19, 46, 77
Foley, Robert, 14
Frazer, J. G., 44, 61
Freud, Sigmund, 43–44, 61, 127
Fxj20, Kenya, 108

Gabillou, France, 83
Ganj Dareh, Iran, 98
Genesis, 2, 122
Genetic studies, 7, 24, 27, 34, 40, 48, 61, 63, 141 n. 2, 142 n. 3
Geney, France, 22
Geophytes, 24
Gesher Beneot Yaakov, Levant, 108
Grand Pastou, France, 80
Grimaldi, Italy, 65, 82
Grotte des Enfants, Italy, 68

Hadza, 116, 117, 119, 126
Harpoon, 12, 76–77
Harvesting of grasses, 93–95
Hau de no sau nee (Irokwa), 131–32
Havel, Václav, 124
Hayonim, Levant, 42–43, 45, 57
Healey Lake, Alaska, 88
Henshilwood, Christopher, 15, 19, 25, 28
Heun, Manfred, 96
Hierarchy, 68–69, 102
Hillman, Gordon, 97
Hohlenstein-Stadel, Germany, 58
Hollow Rock Shelter, South Africa, 25
Homo erectus. See Erectus
Homo sapiens, 2, 12–104, 110

Howieson's Poort, South Africa, 18, 27, 28, 29, 41
Hunting, 14–24, 39, 41, 45–46, 52–55, 69, 73–81, 84–91, 92; by Neandertals, 21–23; psychological effect of, 20, 56–57; strategies, 78–81
Hunting Magic, 42–44, 57–61, 69, 118

Interments, 39, 42, 65–68, 84, 140 n. 1, 141 n. 5
Isaac, Glynn, 113
Istállóskö, Hungary, 49, 55
Isturitz, France, 55

Jameson, Fredric, 128
Japan, 97, 112
Jebel Sahara, Sudan, 84
Jericho, Palestine, 100
Jochim, Michael, 73–74
Jordan Valley, Levant, 96
Jung, Carl, 123, 129

Kapthurin Formation, Kenya, 28
Katanda, Congo, 13, 18
Kebara, Levant, 95
Kelsterbach, Germany, 48
Kent's Cavern, England, 48
"Killed Man," 83
Killing, 81–84, 142 n. 1
Klasies River Mouth, South Africa, 19, 24, 28
Klein, Richard, 14, 18, 19, 25, 40, 51, 53–54, 76, 90, 109, 119
Klein, Sheldon, 41
KNM-ER 1808, 115–16
Kogi, 116
Konigsaue, Germany, 23
Kom Ombo, Egypt, 93
Koobi Fora, Kenya, 108, 115, 119